THE SWORDFISH HUNTERS

The Life, Death, and Future
of *Xiphias gladius*

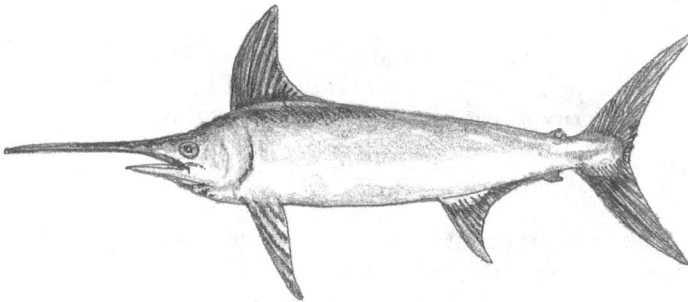

Swordfish *(Xiphias gladius)*

Maximum recorded weight -	1182 lbs
Maximum recorded length -	15 ft

TONY TRH 2018

Dr. Thomas Armbruster

SandyHook SeaLife Foundation (SSF)
www.sandyhooksealife.org
sandyhookpress@aol.com

Cover Art by: Catherine Armbruster

Illustrations by: Tony Troy

ISBN: 978-0-5786-1945-3 (sc)

Library of Congress Control Number: 2019900843

Printed in the United States of America

Lulu Publishing Services rev. date: 04/30/2021

a revision of
The Crimson Broadbill:Commercial Swordfishing the NW Atlantic

Library of Congress Number: 2002093999

ISBN: 1-4010-2907-8
Xlibris

Dedication

This book is dedicated to all those who fight to save the marine environment, for their struggle is immense. It is my sincere hope that this story will inspire every reader to join that effort.

Ocean diver-explorer Jacques-Yves Cousteau summarized it best when he said "For most of history, man has had to fight nature to survive; in this century, he is beginning to realize that, in order to survive, he must protect it."

The ongoing destruction of sea fish, notably big, long-lived, and slow-to-reproduce fish, such as bluefin tuna and sharks, remains sobering. In light of the "pressures on their populations" – a euphemism for killing them early and killing them often – those big fish cannot beget sufficient offspring to survive.

- W. Jeffrey Bolster
The Mortal Sea

THE MORTAL SEA: FISHING THE ATLANTIC IN THE AGE OF SAIL by W. Jeffrey Bolster, Cambridge, Mass.: The Belknap Press of Harvard University Press, Copyright © 2012 by the President and Fellows of Harvard College.

Contents

Section III The Apex Pelagics

Overview of the HMS Fisheries in the Atlantic (pre-2000 - present) 187

Acknowledgments

This acknowledgement extends to the many who have participated in the production of this work, published for peer review in 2002 as *The Crimson Broadbill: Commercial Swordfishing the NW Atlantic,* and affectionately referred to as "Project Broadbill."

First and foremost—to my friend, project manager and number one editor, heartfelt thanks Mary Hamilton, Director of SandyHook SeaLife Foundation (SSF). Many thanks to my sister (Cat) Catherine Armbruster (Authentic Bleu) and SSF's Communications Director, who spent numerous hours in her studio creating the brilliant cover and sectional artwork. And thanks as well to my brother Mike, and my father David, for their support.

Thanks to my great friend Roger Weaver for proofreading, and to my editors Barbara Noe Kennedy (b.ink and former senior editor *National Geographic Travel*), Captain Al Ristori (the *Star-Ledger*), Professor Patricia M. Avolio (English Studies, Fairleigh Dickinson University), and the late Dr. Rudolf G. Arndt (professor of marine science, Stockton University). My gratitude, as well, to Tony Troy (Tony Troy Art) for his beautifully-rendered sketches, to our graphics designer, Rachel Halter (Ocean Copiers), for her creative vision, and to LULU Press for their patience with this project.

Many additional thanks to Captains Keith Larson and John Larson, of Viking Village; Captain Peter Dolan; Nelson Beideman (now deceased) and Terri Beideman (Blue Water Fisherman's Association); Saul Phillips (Export Inc.); J. Tim Hobbs (National Coalition of Marine Conservation); Clean Ocean Action; Maureen Ackerson (Ocean County Chamber of Commerce); Ernie Panacek (general manager, Viking Village); Claire Steimle (NOAA Library Services of Sandy Hook); Dottie Anderson (lead reference librarian, NOAA Central Library); Caroline Woods (NOAA reference librarian); Donovan's Reef Bar (for their inspirational seascape and cold beer); and *always* to the U.S. Coast Guard for their daily heroism. I also want to thank the many other contributors to this project, now too numerous to mention.

And finally, I want to recognize the U.S. commercial fishermen, both men and women, who risk their lives to bring fresh seafood to market. Theirs is a dangerous and highly – regulated industry wherein they constantly strive to achieve a balance of sustainability and profit.

Disclaimer: Due to the highly charged emotional and financial climate of the offshore fisheries, the character surnames, company name, and vessel's name have been changed in order to protect all parties involved in this account. Any reference to the name of past or current commercial longline crews or vessels is purely coincidental.

Foreword

Few things in life are simple and straightforward, but none are more complicated than man's interactions with the natural resources of our oceans. Decades after man landed on the moon, we still know relatively little about oceanic fish and are in the infancy of trying to manage these resources. This effort has led to little more than a slow, steady decline, particularly in oceanic species.

The broadbill swordfish, the focus of this book, long has been highly desired as food and sought after by commercial fishermen wherever found throughout the world. Due to both their great fighting ability and the difficulty involved in baiting them when spotted on the surface, swordfish also have been one of the most desired of all oceanic game fish. There was little problem with this species as long as the commercial fishery was prosecuted by harpooners spotting large swords on the surface. However, when Japanese longlining practices were adopted in the mid-1950s throughout its range, the fishery started crashing within a few years. The net result not only has been a severe decrease in the population, but also the virtual elimination of surface angling in near-shore waters for the species in the Atlantic. Most of the swordfish caught by longline these days aren't large enough to have spawned even once. Anglers now must catch their swords by deep-fishing methods far offshore, and most fish caught are well under one hundred pounds. Fortunately, the new conservation ethic results in most of those smaller sportfishing catches being released.

The difficulties faced by longliners in making their catch is well illustrated here by Thomas Armbruster, even if weather wasn't as great a concern when he made his trip as was the case in *The Perfect Storm*. The key point is that when this trip was made, it still was possible to make a decent catch of large swordfish during the summer in nearby Hudson Canyon. However, in recent years, it's been necessary for longliners to make longer, more dangerous trips to the Grand Banks and beyond, just to have a good shot at a payload of swords.

Making a living from the sea is difficult at best and nearly impossible at worst, but there always will be many fishermen with salt in their veins who would have it no other way. They're to be admired for their tenacity, though those efforts to scratch out a living today can't be allowed to prevent conservation of the species, which is in everyone's interest in the long run. Efforts are under way among conservation groups and in Congress to fund the retirement of an oversize fleet, and there soon may come a day when a much smaller longline fleet will be able to earn a good living while permitting stocks to rebuild.

There have been some hopeful signs since the National Oceanic and Atmospheric Administration's National Marine Fisheries Service prohibited longlining in areas frequented by juvenile swordfish. The swordfish population may have stabilized. Anglers fishing at night in the canyons once again have a reasonable shot at hooking one, while daytime drops in great depths off the Florida Keys have been so productive that my son was able to catch a swordfish on his first attempt out of Bud N' Mary's Marina, Islamorada.

This story brings both the romance of longlining and its hardships into focus along with the need for conservation of one of the sea's greatest treasures. You'll also learn about the many other creatures that interact with longliners and be able to appreciate the type of person it takes to endure weeks of hard labor and little sleep for no guaranteed return at all. It's a fascinating story, and Thomas lays it out with clarity and uncompromising honesty.

Al Ristori

Salt Water Sportsman magazine, regional editor

The Fisherman Magazine, conservational editor

Star-Ledger newspaper, saltwater editor

Preface

This story originated as a shipboard journal/logbook written during the 1980s. Although I was a scatterbrained, impetuous college student, my advisor had, with some trepidation, agreed to award extra credit if I completed a summer work-study program. All I had to do was find a job and document the catch of the fishing vessel while at sea.

Like many others of my generation I was fascinated with the ocean from early childhood. But it was our summer vacations on Long Beach Island, a barrier island off the New Jersey coast, that really 'hooked' me. Afternoons often were spent catching crabs from Barnegat Bay and then racing them on the deck of our beach house, after which 'losers' were then cooked in steaming Old Bay Seasoning, as were the equally- unfortunate 'winners.' I especially was drawn to the marina, the fishermen, and their boats. As I grew older I'd go deep-sea fishing or just hang out around the incoming boats, watching the crew unload the day's catch and listening to the fishermen talk. That's where I heard my best "fish stories," which I would relate later at the dinner table, some of which actually were true, some embellished a bit.

But it was the 1975 film, *Jaws,* that drove me toward a marine biology career. Matt Hooper, the intellectual and sometimes pompous marine scientist (from famed Woods Hole Oceanographic Institution) inspired me with his intelligence, confidence, and passion for sharks, while his knowledge of such a fascinating subject as the great white shark electrified me. I wanted to emulate this free-spirited scientist who possessed little-known information and keen insight. I watched the movie six or seven times, just to immerse myself in the role.

As a teenager I visualized adventures on the *Calypso,* the shipboard home of legendary ocean diver and documentary film maker, Jacques Cousteau. I read every book I could find on marine biology and studied the mighty great white shark, giant bluefin tuna, and fabled giant squid. I memorized every species of North American fish and delighted as classmates quizzed me. This was to be my life, a professor of ichthyology (fish), living on the high seas and studying sharks.

Following graduation, I had decided to stay in state, so I applied to the most accessible college offering a marine biology program. Naturally if the school was on the water, so much the better, and I chose what is now Stockton University. But being an average high school student made college admission tenuous at best. Only through a solid interview and the presence of my intimidating father did I gain admission as a marine science major. The college, a small commuter school located in the quiet town of Pomona, in the beautiful New Jersey Pinelands, provided an ideal setting to relate with professors individually. In addition, the marine science program was full of dreamy National Geographic "wannabes," self included. We envisioned a romantic career sailing the world aboard the *Calypso* while diving the seven seas. We had no clue that the majority of marine scientists lead a rather threadbare existence.

In the spring of sophomore year, a friend and marine science classmate, David, planned a summer of research with the National Marine Fisheries Service, the branch of the government dealing with coastal fisheries conservation. He was assigned to work in Alaska aboard a foreign fishing vessel, getting paid to monitor compliance with international rules and regulations. I, on the other hand, planned to work the summer in a seafood restaurant—as a dishwasher. David suggested that I couple an independent college elective with commercial fishing work, and my plans changed dramatically. The idea of summer work combined with research on the open ocean was thrilling, and I arranged to meet with my college adviser the following day.

My advisor was a strapping six-foot-five, 260-pound marine biologist from Cornell University, an imposing man with a booming voice and tough manner, but a strong source of knowledge and support. After an hour-long meeting, and much debate, he agreed to a summer research project, if I could find the work. It would entail an at-sea study of the New Jersey commercial longline fisheries, detailed in a written journal, and accompanied by the appropriate laboratory specimens to complement the report. He patiently explained that the chance of finding such unusual work was slim, then wished me well, and curtly dismissed me from his cluttered office. As I was leaving I overheard him mumble, "They're going to eat him alive."

Early the next morning I drove to Long Beach Island, feeling overwhelmed and dejected, the hope of finding gainful employment shrinking with each mile. During the next two weeks I returned, visiting every marina, bait store and hot dog cart, distributing hundreds of flyers. The combination of little experience and a dearth of available fishing jobs meant that my chances of employment on a party boat were slim, and they dropped to "zero" if I expected to work as a commercial fisherman. One of my last stops was the historic Viking Village marina, located at Nineteenth Street in Barnegat Light, New Jersey. I placed my number on the bulletin board and headed home, unaware that this one action would change my life forever.

SECTION I

ALL GREENHORNS ON DECK

General circulatory system of the North Atlantic

GREENLAND

ICELAND

NORWAY

KAP FARVAL

ENGLAND

EUROPE

NORTH
AMERICA

SPAIN

BERMUDA

AZORES

Sargasso Sea area
(very little current)

CANARY IS

KEY:

CARIBBEAN SEA

C. VERDE IS

Cold Labrador

SOUTH AMERICA

Warm Gulf Stream

TONY TROY 2018

GULF STREAM East Coast from Nova Scotia to Cape Hatteras

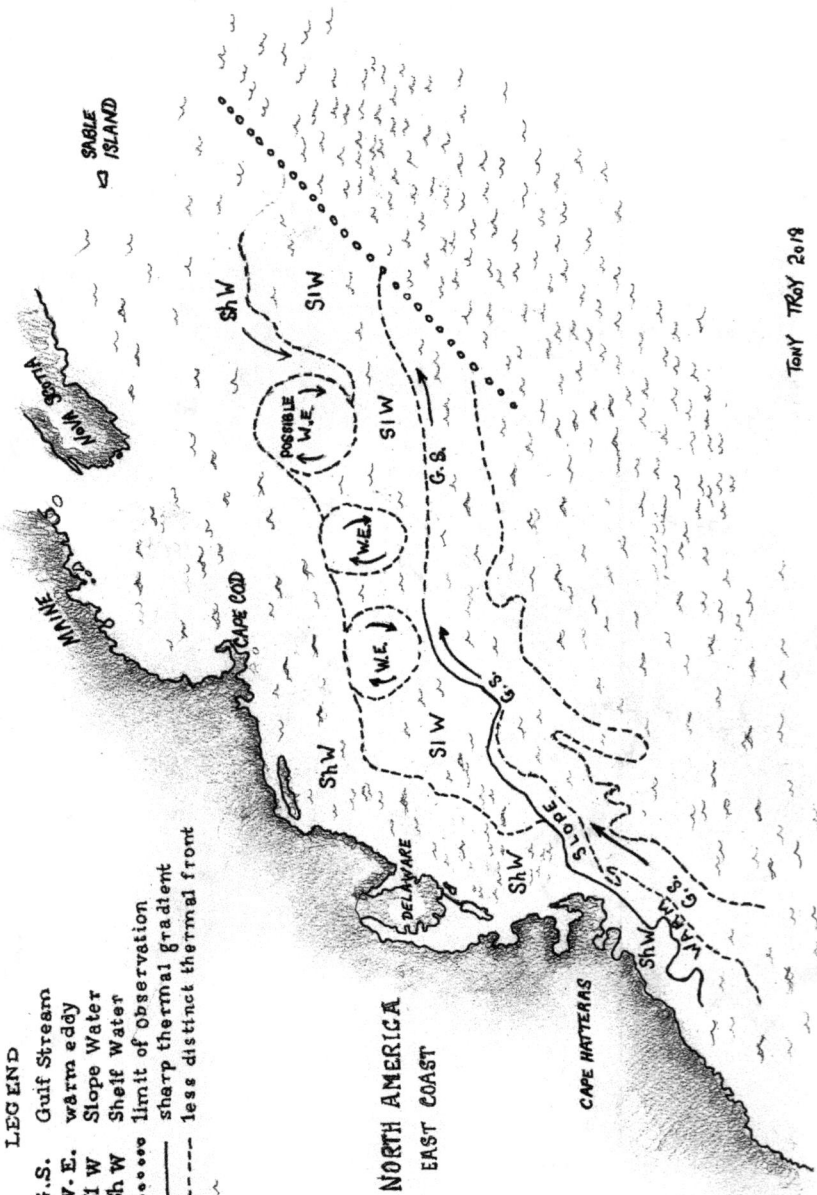

LEGEND

G.S.   Gulf Stream
W.E.   warm eddy
Sl W   Slope Water
Sh W   Shelf Water
∘∘∘∘∘∘ limit of observation
———   sharp thermal gradient
- - - - less distinct thermal front

NORTH AMERICA
EAST COAST

SABLE ISLAND

NOVA SCOTIA

MAINE

CAPE COD

DELAWARE

CAPE HATTERAS

Sh W

Sl W

POSSIBLE W.E.

W.E.

W.E.

G.S.

SLOPE

TONY TROY 2018

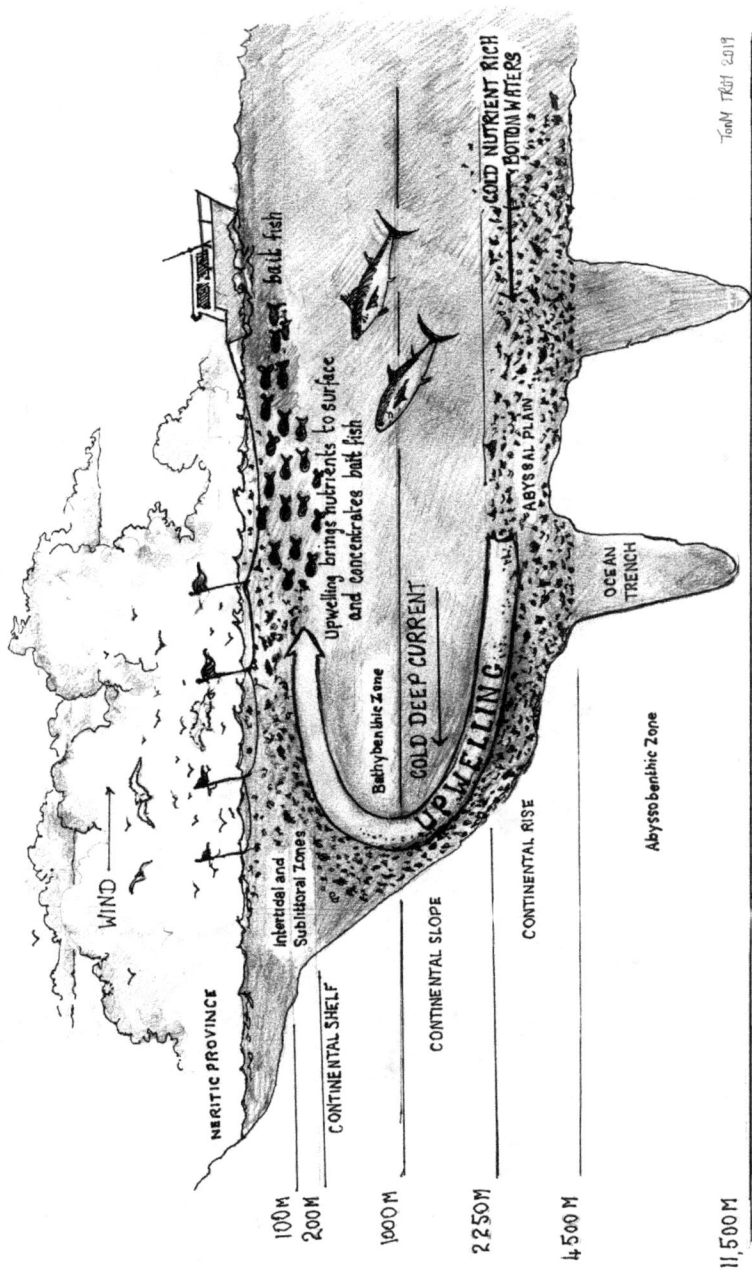

CROSS SECTION DIAGRAM of CANYON STRUCTURE with UPWELLING CURRENTS

NERITIC PROVINCE

Intertidal and Sublittoral Zones

CONTINENTAL SHELF

CONTINENTAL SLOPE

CONTINENTAL RISE

Abyssobenthic Zone

Bathybenthic Zone

COLD DEEP CURRENT

UPWELLING

ABYSSAL PLAIN

OCEAN TRENCH

WIND

bait fish

Upwelling brings nutrients to surface and concentrates bait fish

COLD NUTRIENT RICH BOTTOM WATERS

100 M
200 M
1000 M
2250 M
4500 M
11,500 M

TOM TRPI 2019

States of swordfish growth
(Xiphias gladius)

Swordfish larva,   12mm long

Young Swordfish,   37mm long.

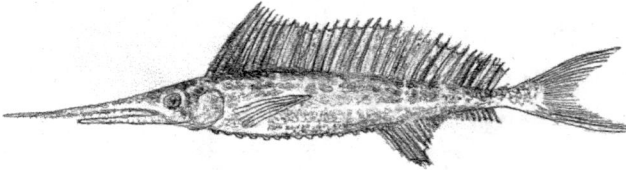

Young Swordfish,   12 inches long are miniature barbwire fortresses playing a defensive role in a struggle for
existence.

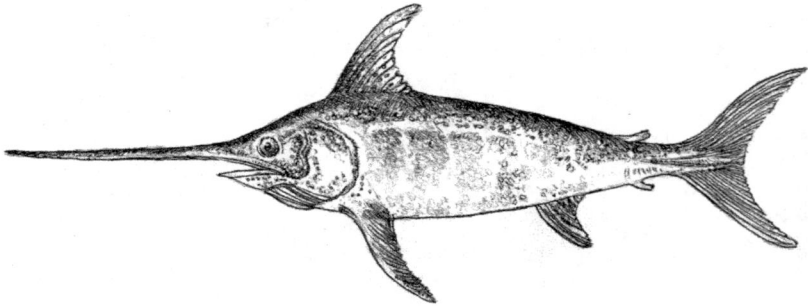

Adult swordfish.

Tony Tray 2018

SWORDFISH

(Xiphias gladius)

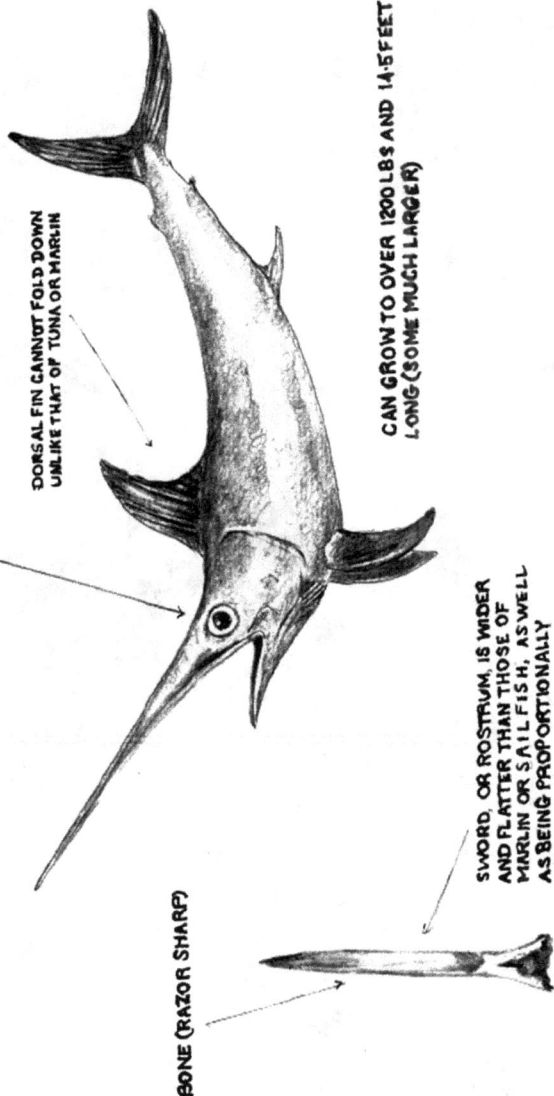

CRESCENT TAIL IS DRIVER FOR SWIMMING SPEED

DORSAL FIN CANNOT FOLD DOWN UNLIKE THAT OF TUNA OR MARLIN

CAN GROW TO OVER 1200 LBS AND 14.5 FEET LONG (SOME MUCH LARGER)

LARGE EYE FOR DEEP SEA VISION > 2000' DOWN

SWORD, OR ROSTRUM, IS WIDER AND FLATTER THAN THOSE OF MARLIN OR SAILFISH, AS WELL AS BEING PROPORTIONALLY LONGER.

BONE (RAZOR SHARP)

TONY TROY 2019

SKELETAL ANATOMY

(Adult Swordfish)

NATURAL ENEMIES OF THE BROADBILL SWORDFISH

MAN & MAKO SHARK

CHAPTER 1
Xiphias gladius

The solitary and elusive swordfish (*Xiphias gladius*), also known as broadbill, velhuella, or shutome (in Hawaii) is a migratory marine carnivore that is distributed throughout the Atlantic, Pacific, and Indian Oceans, and the Mediterranean Sea. An ultimate worldwide predator, the swordfish is grouped among the fastest and largest apex animals in the ocean, a category titled "highly migratory species" or HMS, which also includes sailfish, marlin, spearfish, tunas, and some sharks. While *Xiphias gladius* shares many characteristics with her billfish relatives, the broadbill is distinguished by a rigid, curved dorsal fin, azure-blue eyes the size of tennis balls, and a black-brown body lacking any of the characteristic stripes or spots seen in other billfish.

Swordfish prefer warm, open ocean waters where surface temperatures range between sixty and eight-five degrees Fahrenheit but they can adapt to great depths and changes in water conditions. Some have been caught off Nova Scotia, where ocean temperatures fall into the low forties.

Swordfish have been reported to reach fifteen feet and weigh-in at 1,500 pounds. In *Natural History of the Fishes of Massachusetts* (1833) we read: "On the coast of Brazil, the sword-fish attains its greatest dimensions, being found twenty five feet long." Unfortunately, we do not know if this information was scientifically documented or if it is pure speculation. We do know that a modern world record was achieved in 1953, when a fish weighing 1,182 pounds was taken off the coast of Chile. However, the average fish caught today is under 150 pounds and measures approximately five feet in length.

Swordfish were named for their bill, the stiff, sharp-edged, flattened bone that can grow to one-third the fish's total length and is used as a feeding tool that enables the fish to slash and incapacitate squid and schools of mackerel or herring. Swordfish have been observed engulfing smaller prey. Their feeding method of lateral slashing indicates a tendency to attack vertically while the sword is diving or rising from the deep. In 1974, X-rayed samples of swordfish stomach contents revealed that much of the prey fish had been slashed, resulting in broken backbones.

The reason for this style of feeding is obvious. The high swimming speed attained by the predator requires significant energy. By using its bill as a sharp, slashing tool the swordfish can immobilize a whole school of fish and engulf them in pieces, making it unnecessary to chase whole fish. An analogy would be that of a person hunting geese with a shotgun. The blast would injure or kill more birds in a flock than attempting to hit one bird with a single shot, thereby minimizing effort, and preserving the hunter's energy.

The swordfish is also an opportunistic feeder, analogous to an oceanic terminator, and visual acuity is one of the many physiological adaptations that make this predator so successful. Large, azure-blue eyes and a sophisticated retina allow the fish to see at depths of more than 1,000 feet, while a special heat-generating system located near the brain keeps the eyes warmer than the ambient water by as much as ten degrees. These capabilities enable the sight-dependent hunter to discern miniscule changes in light as the animal darts in and out of varying temperatures and ocean depths. This same system also heats the brain, serving to reduce fluctuations that might otherwise impair hunting behavior.

But the broadbill's greatest strength is her sheer speed. Considered one of the fastest fish in the ocean, with a swimming speed estimated at nearly eighty miles per hour, made possible by an aerodynamic anatomy, temperature-regulated eyes and brain, and lubricating glands in the head that serve to reduce drag.

Their ability to dive to 3,000-foot depths in a short period of time, swim at speeds of more than seventy miles per hour, and use its bill as a sharp, slashing feeding tool makes escape for most prey difficult, if not impossible. The fish generally remain offshore until dusk, then swim inshore to hunt, moving toward the surface to feed on the squid, mackerel, red hake and cod as they, in turn, feast on zooplankton, the microscopic animals that move toward warmer surface water at sunset. But swordfish will take whatever food presents itself. Everything from birds, to a four-foot shark, to bottom-dwelling octopus and tilefish have been discovered in their stomachs.

Although the broadbill's range is worldwide, the western Atlantic population relies on two massive currents (found by fishermen in the 1960s on satellite photos), namely the Labrador Current and the Gulf Stream. Due to their opposing flows and significant temperature differences, these waters have become an optimal feeding ground for millions of fish. An inshore current, the cold water of the Labrador Current flows from the Arctic Ocean, moving south along the east coast of Canada, near Nova Scotia, where it meets the warm northward-moving waters of the Gulf Stream. Due to this natural confluence, the major fishing grounds for the US Atlantic swordboat fleet have historically been along the southern edge of the Continental Shelf, off the coast of Florida, and as far north as the Grand Banks of Newfoundland.

The peak spawning period for the Northwest Atlantic broadbill is during the late fall and early winter but larvae found throughout the year suggest longer periods. The three major spawning areas are the Straits of Florida (Atlantic Ocean), the Straits of Yucatan in the Gulf of Mexico, and near the Lesser Antilles (southern Caribbean Sea). As in most fish, fertilization is external, a very effective method.

The swordfish, like other Highly Migratory Species, has a very high fecundity rate, or ability to produce an abundant number of eggs. The capture of a 440-pound female revealed 6.2 million eggs, but numbers ranging from 3 million to 16 million eggs per spawning are possible. It is estimated that the survival of any one egg to adulthood is one in several million, meaning that only two to three fish from each spawning will live to adulthood.

Age at first spawning is about five years, when males have grown close to thirty-nine inches and weigh just under fifty pounds. Females do not reach reproductive age until they have attained a whopping 161.3 pounds and measure at least thirty inches. Some researchers suggest, however, that fish of less than fifty inches are still immature.

Once hatched, the tiny larva change dramatically. By the time they reach twelve inches they have developed a barbwire appearance, a short snout, sharp teeth, and a scaly body. Still not resembling the adults, they continue to change form rapidly as they age, eventually losing their teeth as the dorsal fin narrows. The advantage of quickly achieving size is necessary for survival. A first-year fish may weigh only nine pounds while a four-year old could weigh 150 pounds and is considered mature. Males are smaller, perhaps due to their tendency to remain in warmer Gulf Stream currents containing less nutrient-rich, but deeper, waters. And it is perhaps lower caloric intake that creates such disparity in the size of the sexes. It has also been suggested that the genders hunt in different waters, thereby reducing competition for food and allowing the females to meet their higher nutritional needs for egg laying.

❦

The swordfish was virtually unknown in colonial America, and its commercial value as a game fish, or food fish, was not recognized until the nineteenth century. Although Nantucket whalers reported seeing basking swordfish, there was no interest in capturing the fish for market until the early 1800s, when a sword-fishery began to develop in New England. But it was

New Jersey that gained credit as one of the first states to consume the unusual fish. In 1817, the *American Monthly* magazine reported that a swordfish was taken by harpoon off Sandy Hook, New Jersey, and sold at a New York market for a respectable price. Surprisingly, people were willing to purchase and sample this mystery meat without fear of consequence. The *Barnstable Patriot* of June 30, 1841, mentions the first recorded acknowledgment of the palatability of swordfish: "Fishermen of Martha's Vineyard capture a small amount of swordfish, pickling approximately two hundred pounds a year."

By the 1860s, only twenty years later, the same paper reports: "Thirty vessels are fifteen miles south and east of No Man's Land or sixty miles out of New Bedford and same distance from Nantucket. The season extends from June to September. The fish generally weigh 400-500 pounds and are from ten to twelve feet long. They are sold in New York." This fleet probably was based out of Martha's Vineyard, although New Bedford appeared to be a lesser center of activity. Sword steaks sold at four cents per pound, and salted swordfish went for six dollars a barrel.

But the availability of fresh swordfish, at this time, was severely limited due to the lack of proper refrigeration. Instead, it was necessary to preserve the meat—at least two-thirds of the catch—through other methods such as pickling with vinegar or the more common method of salting, a process that cured the steaks with heavy doses of rock salt, thereby removing liquid and drying the fish to a hard, jerky-like texture. Dry-salted fish also weighed less than fresh, so it was easier to transport. When needed, the fish could be rehydrated for cooking. Literature from this era referred to salted swordfish as a delicacy that was superior to other species of fish in taste and quality.

Salted sword could be preserved efficiently, and transported in a cured state, for up to ten months. By 1890, ice was kept onboard vessels to preserve the catch, turning fisheries into fresh seafood markets. Dropped into the hold in three-hundred-pound blocks, thirty to forty tons of ice were necessary for a one-month voyage; blocks were preferred over chipped ice because they held up longer and presumably saved time when packing the huge body cavities. But there was a downside as well. Crew members were sometimes seriously injured during the loading process before the ship had even left the dock.

During the 19th century, harpooning was the fishing technique used to capture the broadbill. Because the fish hunts in cold, mid-to-deep, nutrient-rich waters, it needs to rest in the warmer surface water to aid digestion and invigorate muscles. So with a belly full of squid, the fish would rise to the ocean surface to bask, putting it in a lethargic state. However, the behavior had its drawbacks. The super predator—man—recognized this as an opportunity to kill the great fish.

The hunt would begin when the surface of the ocean had warmed. Sailors, called 'spotters,' watched from the crow's nest, a small structure at the top of the main mast, for the dark, crescent dorsal with the characteristic trailing tailfin. Usually younger men with good eyesight, they'd wear long black visors to protect their faces from the sun. Invariably, however, after several hours of scanning the water, they'd have to rest, placing cool, wet tea leaves over their eyes to relieve the strain. When a fish was sighted, they would yell "finner," replacing the "thar she blows" cry typically identifying whales. The harpooner would then take his position at the bow to wait for the best opportunity.

The harpooner may have been the most important person on board, as he had trained from boyhood to hit targets tossed into the sea. He waited on a raised platform in the bow, known as the perch or pulpit, as the ship glided toward the resting fish. Using a harpoon or spear-like tool with a detachable harpoon head and nine hundred feet of line, the end of which was attached to a barrel, the harpooner aimed for the fish's head to prevent destruction of the valuable flank meat. Once the six-inch steel harpoon penetrated its flesh, the fish reportedly convulsed in pain and would instinctively dive toward the safety of the cold deep water. Blood loss, combined with the effort of straining against the floating barrel, rendered exhausted fish ripe for the kill, often within twenty minutes.

The battle usually ended when the mortally injured fish either was mutilated by sharks or captured. Several men in small wooden dories would haul the animal to the surface, then beat it to death or use a shotgun to the head. Although man was the victor in nearly all cases, the pugnacious sword sometimes had her[1] revenge.[2] While they showed no distrust of larger sailing

---

[1] Larger swords are female and were most often hunted rather than the smaller males.

[2] The broadbill has a reputation for its irritable disposition and appears "to be at open war with whatever moves in the same liquid element," according to *Natural History of the Fishes of Massachusetts.* It is infamous for using its sword to attack boats, a fact recorded as far back as the 1800s. In 1871, the *Red Hot,* a yacht on a pleasure cruise, was sunk after a broadbill rammed a large hole in the hull, a reportedly unprovoked attack. Another story from the same era tells of a sword driving its entire length into the solid wood hull of a ship, killing the animal immediately. More recently, the Woods Hole submersible, *Alvin,* was working in two thousand feet of water along the Blake Plateau, off the coast of Georgia, when an aggravated sword rammed its bill into the soft gasket

vessels, presumably recognizing them as harmless whales, this was not the case with small dories, which caused the fish to sound. To this day, there are documented reports of swords ramming and sinking small dories, and many lives have been lost before the mother ship could save the dory crew.

Why would a harmless boat trigger such outrage? One theory suggests that the animal's pain and anger, coupled with extensive brain trauma, sets off the attack. Numerous stories relate how a harpooned fish visually followed the line back to the dory. It then attacked the boat, apparently associating the line and boat with the source of its pain. Another theory is that a small dory may resemble a shark, a feared enemy of the broadbill for millions of years. But one of the most amazing illustrations of the broadbill's ferocity occurred in May 2015 in Hawaii. A fishing boat captain noticed a juvenile swordfish swimming alongside his boat and, with a gut full of gin, the man decided to spear the sixty-pound fish. He jumped into the water and hit the animal, but the fish reacted by attacking and skewering the captain through the chest with its three-foot knifelike bill. Reports indicate that the unfortunate captain was speared directly in the heart and died within minutes of the event. While seldom found in shallow water, it is likely that the fish was either injured or sick.[3]

<hr />

material of a viewing port. The swordfish had wedged itself so tightly that it could not be freed and was reportedly taken aboard the ship and eaten for dinner.

[3] In addition to an irascible disposition, the broadbill also is curious, particularly about floating objects. A story, related to me by a close friend, describes such an encounter

The modern commercial swordfishery began in the Northwest Atlantic in the late 1950s with Japanese fishermen, who were the first to use thousands of baited hooks strung along miles of lines to catch swordfish, billfish, and tuna. As illustrated in Charles Dana Gibson's book, *The Broadbill Swordfishery of the Northwest Atlantic*, the Japanese fished the Atlantic Ocean in very logical and aggressive search patterns, crisscrossing the latitudes. By the late 1960s, word had spread of their success, and the U.S. Northwestern Atlantic sword-fishery was born.[4]

Prior to the 1970s, the earliest commercial U.S. swordfishing grounds were found mainly in the Northwest Atlantic. However, during the '70s, off of Florida's east coast, the tilefishing industry was developing, targeting the two-hundred-meter curve along the Gulf Stream. Tile fishermen were reporting reels being stripped of lines by a large unseen predator. According

---

in the winter of 1994 in the Gulf of Mexico. The crew was sportfishing for shark in the middle of the Gulf. The sun had set, and the powerful boat lights over the shimmering 80-degree water had attracted thousands of squid. As he stood watching them, a huge shadow darted near the boat, then moved toward the surface in a determined and exploratory manner. With all the squid in the water, he assumed it was hunting. Suddenly, the billfish broke the surface with her sword, just five feet away. With a single stroke of her tail, the six-foot fish quietly glided to the boat, unaware of being watched. She tapped her bill on the boat. Getting no reaction, she glided the length of the boat, continuing to observe it. Drifting past the bow, she must have felt the vibrations of the startled crew or seen them onboard and, with a powerful stroke of her tail, she angled for the deep and vanished.

[4] At the same time that the Japanese fishery began, the Norwegians also had started a western Atlantic commercial longline fishery for the prized porbeagle shark. However, within a decade, this shark fishery became commercially extinct due to overfishing.

to reports, witnesses watched as swordfish regularly followed the hooked tilefish to the surface, one huge specimen actually tracking her prey right to the boat. Such observations led to the development of a commercial swordfishery in the Florida Straits.

In its developing years, longlining was not without its problems. For example, sharks and other predators decimated the gear. Usually the first one to two days didn't see much damage, but the crew learned not to set in the same location by the third day. In addition, their migratory and predatory behaviors were poorly understood, making it more difficult to find swordfish. Boats also lacked the modern-day weather information, provided by satellites, to read water temperature gradients necessary to find fish.

Under a steep learning curve, the U.S. fishery has evolved into a highly efficient method of finding and catching the broadbill. But there is a price. It's brutally hard work in a hostile environment, requiring tremendous physical and financial resources, and few giant fish remain in the Atlantic, making it necessary for the U.S. swordboat fleet to expand its fishing theater into the Pacific and the Straits of Florida. Although the East Coast swordboat fleet fishes the Northwest Atlantic during the summer months, the majority sail to the Caribbean during the winter, following the highly migratory patterns of their elusive prey. It is because of these patterns that the entire fleet constantly must adjust in pursuit of the broadbill.

More than other cold-blooded fishes, swordfish have managed to survive for millions of years because of their power, size, speed, and weaponry, all due to a unique anatomy. The animal's internal heating system, called rete mirabile (also found in tuna and some sharks), warms muscles to ten degrees above surrounding water temperature, facilitating better muscle contraction and rapid digestion. Also present in the brain, this unique system affords the animal higher sensory functioning for hunting and evading. In this regard, the broadbill is more like a warm-blooded animal.

Clearly, the swordfish has earned its title of "gladiator of the seas." Despite its attributes, however, the fish has two great flaws that have caused the species to decline over the past two hundred years: the fish cannot evade its top predator, and it tastes great on the grill.

By 2000, these two simple facts resulted in a crisis. Sword-fisheries— and those of other highly migratory species (HMS)—plummeted to a shadow of their previous levels. The average annual HMS catch had shown increases throughout the 1970s and 1980s, with a whopping eight thousand metric tons caught annually—up from five thousand metric tons of fish caught annually in the 1960s. And this number increased even further in ensuing years due to advanced science, technology, and the pressure to produce more fish. But statistics in 2000 indicated the obvious; we were overfishing a declining population. And as serious as that is in itself, it was only one of the many complex, intertwining problems the ocean —and all of its creatures—experience in varying degrees today.

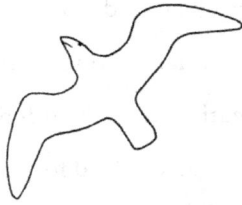

CHAPTER 2
Quint Calls

The call to work on a commercial fishing boat came late on a June evening; the unfamiliar voice, gruff and terse, was difficult to hear due to the pounding rain in the background.

"You Tom?"

"Yes."

"I'm Capt'n Coyle with the swordboat longliner *Defiance*. Heard you're lookin' for fishing work."

"Yeah, I am."

"Boat leaves tomorrow at 8:00 a.m. from the Nineteenth Street Marina. Bring plenty of clothes 'cause we'll be out awhile."

Click.

In the vacuum of silence left behind, I pictured the face of Capt'n Quint, the epitome of the rugged seaman who famously hunted for *Jaws*. I tried to shake the image of a weatherworn, bearded, angry-sea-wolf of a man with salt in his veins and a healthy hatred for landlubbers. I knew captains could make or knowingly break a life at sea, and I prayed this one would be merciful. With that, I headed for bed, knowing morning would come too early.

But thoughts of deep-sea adventures prevented restful sleep. The rain continued as I tossed and turned, the night dragging on, while I imagined my first oceanic voyage as one of the "big boys." I would soon be a fisherman on a commercial swordfish longliner. I wondered what the crew would be like, how long we'd probably be at sea, the size of the fish we'd catch, and how the men would relate to a green deckhand. My apprehension grew stronger when I realized that there was no turning back.

Dawn finally broke under heavy clouds. It was 5:00 a.m., and the house remained quiet as I brewed a pot of coffee. Packing was a chore in itself. The list included five pairs of jeans, ten T-shirts, three sweatshirts, rubber boots, sneakers, a heavy denim jacket, oilskins (rain gear), twenty pairs of socks, long johns, two toothbrushes, and all the underwear I owned. Our cat, Bandit, proved to be the one family member loyal enough to see me off at this ungodly hour of the morning. He sat on my lap and purred, seeming to know that I would be away from home for quite a while. I had to admit that the cat was much wiser than me; he at least had the common sense to remain warm and dry ashore.

By 6:15 a.m. I was ready to take to the road. I threw on a best-loved pair of old Levi's, my favorite white "Save the Humans" T-shirt (with a picture of an imploring whale as the speaker), and climbed into my 1973 Super Beetle. The engine roared to life with its characteristic sewing machine buzz. I opened the sunroof and turned up the radio to the sound of "Going Mobile" by The Who. A good omen, I thought, as I shifted to first gear, and drove into the experience of a lifetime.

<hr />

The main road to Long Beach Island is Route 72 East, that famed and still-desolate stretch leading some forty miles past Medford, through the New Jersey Pine Barrens. It eventually reaches the eastern coastline by way of a causeway bridge, ending on an eighteen-mile-long barrier island known locally as LBI.

Fifty years before paved roads and the automobile, trains were making the trip in an hour and a half, running west to east across the state. Not until 1936 was a road paved through the lonely Pinelands, and at times the journey itself provided more adventure than the destination. Motorists wearing goggles and dusters braved mechanical breakdowns and occasional highway robberies to make the four-hour trip from Camden to Beach Haven, crossing over the new toll causeway. The vast Pinelands were so remote and sparsely populated back then, as they remain today, that thieves seldom were caught.

My destination, Barnegat Light, located at the island's north end, is a popular tourist attraction in summer and has played a central role in the history of the island. Barnegat is a derivative of the Dutch word *Barendegat*, meaning "inlet breakers." Historically, the treacherous shoals near the island had destroyed many ships, demonstrating need for a lighthouse, and construction on Barnegat Light began in 1834. The original forty-foot structure stood firm for many years but was considered inadequate and ultimately collapsed. In 1857, Major General George Meade oversaw the construction of a new lighthouse, which was placed three hundred feet farther inland. The 163-foot tower, made of brick and iron and topped by a first-order Fresnel lens, served as an invaluable aid to navigation for inbound ships sailing into New York Harbor until service was deactivated in 1944.

Barnegat Light was one of many lighthouses built along the East Coast during the early days of sail, each with a uniquely coded sequence of light flashes to identify its location at night. The sequence was well-known to mariners. Old Barney's code flashed every ten seconds at each compass point. On a clear day, sailors could see the lighthouse from thirty miles out at sea, identifying it by its day mark or unique color pattern of red on top and white below.

Near the lighthouse, Viking Village was settled in the 1920s when the Scandinavians first chose the island as their home. The island's sheltered harbors and rough inlets reminded these settlers of their homeland and the riches of seafood, both in variety and quantity, available there. They fished throughout the year in all kinds of weather, using pound nets and fish traps along the local beaches and wooden surfboats to collect the catch, frequently paying for their catch with their lives. Later, in the 1940s, they upgraded to lobster fishing and dragging for ground fish. With the introduction of longlining in the late 1960s, the village transformed into a hub of commercial fishing, and it remains so to this day.

The original Norwegian fishing shacks still stand along a dirt road at the marina's southern end. Made of classic shaker shingle, these tiny buildings had been converted to stores offering fishing supplies, clothing, and souvenirs.

———(◯◯◯◯)———

I crossed the causeway bridge and headed north on Long Beach Island, the familiar sight of the Barnegat Lighthouse just ahead. This end of the island now is dotted with exorbitantly priced beach houses, beautiful and freshly painted mansions representing New Jersey's wealthier citizens and the summer playground of many celebrities. The parking lot at the Nineteenth Street Marina was a study in contrast with a couple of shiny-new BMWs sitting between the local fishermen's rusting-out pickup trucks. I found a spot along the periphery, grabbed my gear and headed toward the water. This obviously was a working fishing port. A huge freezer, placed conveniently close to a fish market, conspicuously warehoused the commercial catch, its faded blue walls depicting numerous fish species above the words "Welcome to Viking Village." It released a not-so-subtle fish smell into the salt air.

Men in tattered jeans, sun-faded T-shirts, and well-worn fishing hats tended a variety of boats tied at the docks. The intense activity bombarded me with both new and familiar sounds: hammers on metal, diesel engines revving, swearing. The distinctive National Oceanic and Atmospheric Administration (NOAA) marine band radio blasted weather forecasts from a moored sportfishing boat.

Several types of boats, each with a distinct purpose, style, and crew were in port. While the commercial boats remained berthed alone at the southern edge of the marina, the 'party' boats and charters were tied close to the fish market.

Party boats cater to summer tourists lured by the signs for 'deep sea fishing' strung out along the coast, inviting them to catch some of New Jersey's most edible sport fish like the summer flounder and bluefish. They offer affordable day trips for twenty to thirty fishermen, while the charters, a separate entity, provide a private fishing experience for four to six fishermen on full day or overnight trips.

I noticed that the bulletin board next to the dockmaster's office was empty. This is where I had posted my resume only weeks earlier, before I knew how commercial crewmen were hired. Affectionately called the "Pick a Dick" board, the name aptly implies that nearly any male applicant old enough to talk, and still young enough to walk, could be hired once his name was posted. Most longline captains had steady crews and dreaded a rapid turnover that would result in the need for new and untested crewmen. Another fact, at least according to captains, was that redheads get seasick. Although I didn't fit the look of a weathered seaman, at least I wasn't a redhead. I was a hired crewman about to go to sea, and it was time to find the *Defiance*.

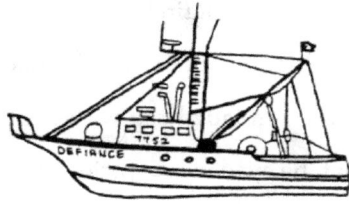

CHAPTER 3
The *Defiance*

Six commercial vessels were moored at the dock. I studied each one but was unsure which was a sword boat. I asked a nearby deckhand; irritated by my ignorance, he called back sarcastically, "Look for boats holding them fish with the long beaks. They'd be called swordfish." As he disappeared below deck to the sound of raucous laughter, three of his fellow crewmen emerged to leer at me. It would not be the last time I'd be laughed at. I moved on, trying to identify the name of each vessel. The last boat at the far end of the pier was my final option, the only possibility – and it was the *Defiance*. Could this be home for the next month?

Bruce Springsteen's "Thunder Road" blasted from the ship's radio over the sound of hammer to metal below deck. The lyric, "You ain't a beauty," perfectly described this ninety-foot-long rust bucket with chipping paint. Hoping for assurance before I stepped aboard, I sheepishly yelled out, "Hello?" A loud expletive roared from below, the source of which appeared shortly after in the form of a tanned, unshaven man in his mid-forties, grumbling obscenities under his breath. Reminding me of George Clooney's character in *The Perfect Storm*, Captain Russell Coyle stood an even six feet, with a square jaw and salt-and-pepper hair. Dressed in tattered jeans, an oil-smeared T-shirt, and a sun-faded cap with the SeaMist company logo emblazoned on the front, he eyed me suspiciously.

"You're the college boy on the phone yesterday?" he finally said.

"That's me."

"I'm Cap'n Russ Coyle."

We shook hands as he continued his appraisal.

"Store your gear below and grab a cup of coffee," he ordered. "Have a look around while we wait for the boys. Cooters and Jack went for last-minute cigarettes. We'll be under way soon as I fix this damn pipe."

Like it or not, I apparently had the right boat.

The *Defiance* was built in 1968 as a scalloper. The steel-hulled vessel spent her first years fishing the waters off North Carolina, her birthplace. After changing owners several times, she finally was bought by SeaMist Inc. and entirely refitted for swordfishing. The company moved her north in the early 1980s, where she would be within close striking distance of the fertile Northwest swordfish grounds.

She was not what I had expected for a commercial longliner. She was smaller, older, and rustier than I'd imagined, and she did not appear as if she could withstand the rigors of the open ocean. The vessel was hunter green to her gunwales and navy gray on top. Two huge pyrovanes, or outriggers, jutted out fifteen feet from each side of the ship. They stood bolt upright but could be splayed at ninety degrees to the water for stability in heavy seas. A huge steel drum, eight feet across, sat bolted to the rear deck just aft of the wheelhouse—which is exactly what it sounds like, the small cabin that houses the ship's wheel and main controls. The steel drum held fifty miles of spooled nylon ground line, a cable much thicker than piano wire. When fishing, the ground line would be spooled out from the drum into the water, much like fishing line from a reel.

Continuing to explore the boat, I noticed that numerous oil stains, chips, and scratches from years of heavy fishing marked the steel gray deck. At the stern lay many yellow and white lines of various gauges, all coiled and waiting for use, along with five cylindrical beacon buoys. Each of the buoys, three feet tall and painted fire-engine red, had a circular white float attached at the top. They were carefully lashed to the railing so that their four-foot-long antennae jutted unencumbered from their tops. Each of these beacon buoys was fitted with a gauge used to monitor the charge of a six-volt battery required to produce radio-tracking signals; its white, flashing strobe light was designed to prevent loss of gear. Since the buoys were worth several thousand dollars each, the tracking system was paramount to profitability.

As I walked toward the stern, a circular hatch, five feet in diameter, caught my attention. I loosened the cover by turning the attached wheel and lifted it open. Inside hid a massive thirty-by-thirty-foot cave. I had found "the hold." It reminded me of a horse stable, as it was divided into four sections or pens with removable oak doors that could accommodate over forty thousand pounds of fish and five thousand pounds of ice. It was cold, damp, and dark, with only sunlight and a bare bulb for illumination. Five feet of ice, poured through a two-foot funnel, covered the floor.

During the voyage, the hold would be filled with swordfish, which, thanks to the ice, would remain eighty percent frozen for as long as a month at sea. There was no active refrigeration unit in the hold, so only the seal kept the ice frozen and the fish cold. Experienced crewmen can examine the hold with a quick look and estimate the weight of the current catch, the approximation based on the number of pens filled with fish and the height of each pen. The hold now contained only bait, about twenty-five hundred pounds of frozen whole mackerel and squid packed in three-by-two-foot boxes, each fifty-pound box loaded by hand. Also visible in the ice were gallon jugs of milk, six cases of Pepsi, and, much to my dismay, only two measly six-packs of Budweiser. It was apparent that the captain wanted us sober while at sea.

Climbing up from the hold, which had no ladder, required some athletic ability. A crewman had to step on a wooden hold divider, then grab the closest piece of metal and pull himself toward the light of the open hatch. Such a climb involved a good deal of body strength at dock; it would require triple the effort on a tossing sea. As I closed and tightened the hatch wheel to preserve the ice within, I realized I would need to strengthen my upper body to get through this trip.

Toward the bow, a six-foot-tall hatchway led into the galley, sleeping quarters, and engine room. A single bulb surrounded by a gray metal cage provided the only deck light on either side of the hatchway door, but powerful deck lights stood on either side of the wheelhouse to illuminate the ship during long nights of work.

Walking into the galley through the open door, I inhaled a mix of diesel fuel and strong coffee— two smells I would come to always associate with the sea. Well-worn, once-red carpeting covered the floor. To my right stood a six-by-three-foot metal freezer filled with one hundred pounds of meat— the largest steaks I'd ever seen, pork chops in thirty-pound packs, whole chickens, turkeys, and, at the captain's request, spareribs—but no fish. (I later learned that the captain hated eating fish, an aversion he explained as "familiarity breeds contempt," according to the first mate.) Stacked packages of cold cuts and frozen vegetables wedged into every remaining available space.

The nearby refrigerator was well stocked with all the comforts of home including milk, juice, and enough nitrate-laden lunchmeat to cause cancer in thousands of laboratory rats. The cabinets held hearty, calorie-dense, high-energy junk foods like oatmeal cookies, pretzels, and the famed Pepperidge Farm cheddar goldfish. (Apparently the captain was okay with this variety of fish.) Considering my minimal experiences with grocery shopping, I could guess the effort and cost required to pick up this great quantity of food. The remaining cabinets and drawers were full of various mismatched dishes and plastic items, all of which added to the general feel of disarray. What would it be like after a few weeks at sea? Off to the side, a straight wooden ladder appeared to lead directly up to the wheelhouse, the command post of the ship, and the captain's personal territory.

The liquid I poured into the Styrofoam cup looked and tasted more like 10W-40 motor oil than freshly brewed coffee but the hot brew helped to ward off the cool-damp morning air. My stereotypical image of a ready-at-dawn commercial fisherman already was tarnished. My dive watch read 9:05 a.m., and I had fully expected we would be at sea. I was unaware that we were delaying departure until high tide, when the six-foot draft of the *Defiance*'s heavy hull could clear Barnegat Inlet.

Behind me, in the engine room, the three-hundred-horsepower twin diesels whined to life for a pre-sail test amid a plume of thick black smoke. A mix of gas and oil fumes filled the noisy, small, cramped room. Just inside the hatchway door hung the only hearing protection: a single pair of rotting black earmuffs.

I moved toward the bow and back through the galley, where I found the crew's quarters. The space was no more than four bunk beds called "racks," two on the port side and two on starboard, separated by a narrow walkway between. To say this was a tight area would be an understatement. The room smelled moldy and somewhat fishy. The wooden walls bore numerous pinholes marking old and long-forgotten photographs. Wrinkled sheets, pillows, and tattered blankets were stacked on the bunk beds. Most noticeable was the absence of any portholes. Each bed had a small reading light at the head, but very little room to maneuver when lying supine. I soon learned that once at sea, there would be very little time actually spent in the rack. I climbed into the coffin-like top bunk. The accommodations were very different from the spacious cabin I had expected.

I headed back to the galley and climbed the bridge ladder to join the captain. The wheelhouse, or bridge, is the brain of the boat, where its instruments serve to gather information from the surroundings and control the boat's movement. A single, worn leather chair was bolted to the deck in the ten-by-fifteen-foot room, and a huge, spoked, wooden wheel stood in front of the captain's chair.

A small instrument the size of a shoebox, hung above the ship's wheel at eye level. The captain called this the LORAN (for "long-range navigation") and explained how it is used to map out the ship's coordinates through a digital display of latitude and longitude. The alarm-clock-like display allows the captain to set or hold a course and to determine position at sea. The LORAN is accurate to hundreds of meters but has been replaced by the more accurate GPS, or global positioning system, which uses satellites for location in place of ground radar.

I studied each item with care, realizing that every newly-learned skill might contribute in some small way to the success of the voyage. The next key item in the wheelhouse was the radar system, a circular screen on a flat table just in front of the ship's wheel. The screen was angled, enabling it to be seen from the captain's chair. This single piece of equipment was to be the most vital to the safety of our boat, allowing the captain to search for approaching ships in order to prevent a collision. The risk of collision at sea is omnipresent, as even small objects, such as floating timber or trash, can cause a boat to sink. Our proximity to the high-traffic shipping lanes only added to the danger. Captain Coyle instructed me to call him "Russ" before explaining the radar and warning me that I would have to understand the basics before we left dock. I frequently asked questions, trying to memorize every detail of the bridge.

The radio was located adjacent to the LORAN, and it would chatter incessantly throughout the trip. This vital piece of technology contributes a twenty-four-hour NOAA marine forecast and is crucial to preparation for storms at sea. A weather fax rested next to the radar screen, its function to provide incoming storm fronts as a graphically printed diagram. The captain gave a detailed explanation of how to use every piece of equipment and followed each description with, "Got it?" I feigned complete understanding.

Several old and tattered *National Fisherman* magazines lay on the chart table to the right of the captain's chair, where I spotted a shiny new magazine of ill repute, all within easy reach. A small bulb hung over the table to illuminate the piles of worn charts buried beneath packs of Camel cigarettes. A small wooden placard nailed to the chart table and stamped with the company insignia read, "We lead where others follow." The radio squawked a marine forecast as I moved back toward the galley, continuing to investigate this new and mysterious steel home.

I proceeded down the wheelhouse ladder to the deck and toward the forward hold. I could see the captain watching me from the wheelhouse, so I pretended to look the role of a seasoned sea dog, pulling on lines and looking self-assured. At first he smiled, and then he laughed.

"Relax college boy!" he said over the PA system. I breathed a sigh of relief. At least he had a sense of humor.

Opening the forward hatch, I climbed down into the deep hold full of long aluminum poles termed "hiflyers." Each measured twelve feet long and sported a red flag at the top, a buoy in the middle, and a weight at the bottom. Adjacent to the hiflyers, ten large plastic reels held hundreds of hooks attached to coiled lines, called gangions. Also in the forward hold I spotted two-foot-long buoys shaped like an orange bullet and attached to several feet of line. They're called "bullet buoys," though I had no idea what each was used for and would have to wait to see how the operation worked. I heard voices above and realized the crew had returned from their errands. With trepidation, I climbed out to meet my new shipmates.

"Cooters," so nicknamed for reasons left to the imagination, met the stereotype of a commercial fisherman. His tanned, muscular physique matched my height at five feet nine, and he sported a large, heart-shaped tattoo on his huge right arm. A freshly lit cigarette dangled from the corner of his mouth. With a confident stride and deep voice, he welcomed me aboard before disappearing into the wheelhouse. I was later to discover that this well-seasoned mate had a seafaring background similar to our captain's.

Jack was the tallest of the crew, standing at six feet three. He was married, an ex-football player-turned-construction worker, and this also was his first trip aboard a commercial boat, having left the tedium of a low-paying construction job. Our conversation progressed as we waited for the final word from the captain; I grew more excited, while Jack appeared to turn a faint shade of green.

Dr. Thomas Armbruster

The last member of the crew was our mascot, a sheepdog–German shepherd mix affectionately named Mutt. Cooters' newly-adopted pet, an unwanted animal that was abused by its former owner, weighed in at seventy pounds. He growled as I approached him, seizing the opportunity to nip at my jeans and taunt me while the crew prepared for departure. Although a great lover of animals, I considered this a bad omen. I hadn't anticipated a maritime adversary with four legs, but I put the thought out of my mind. It was time to go to sea.

SECTION II

FROM GREENHORN TO SWORDFISHERMAN

NORTH ATLANTIC CANYONS

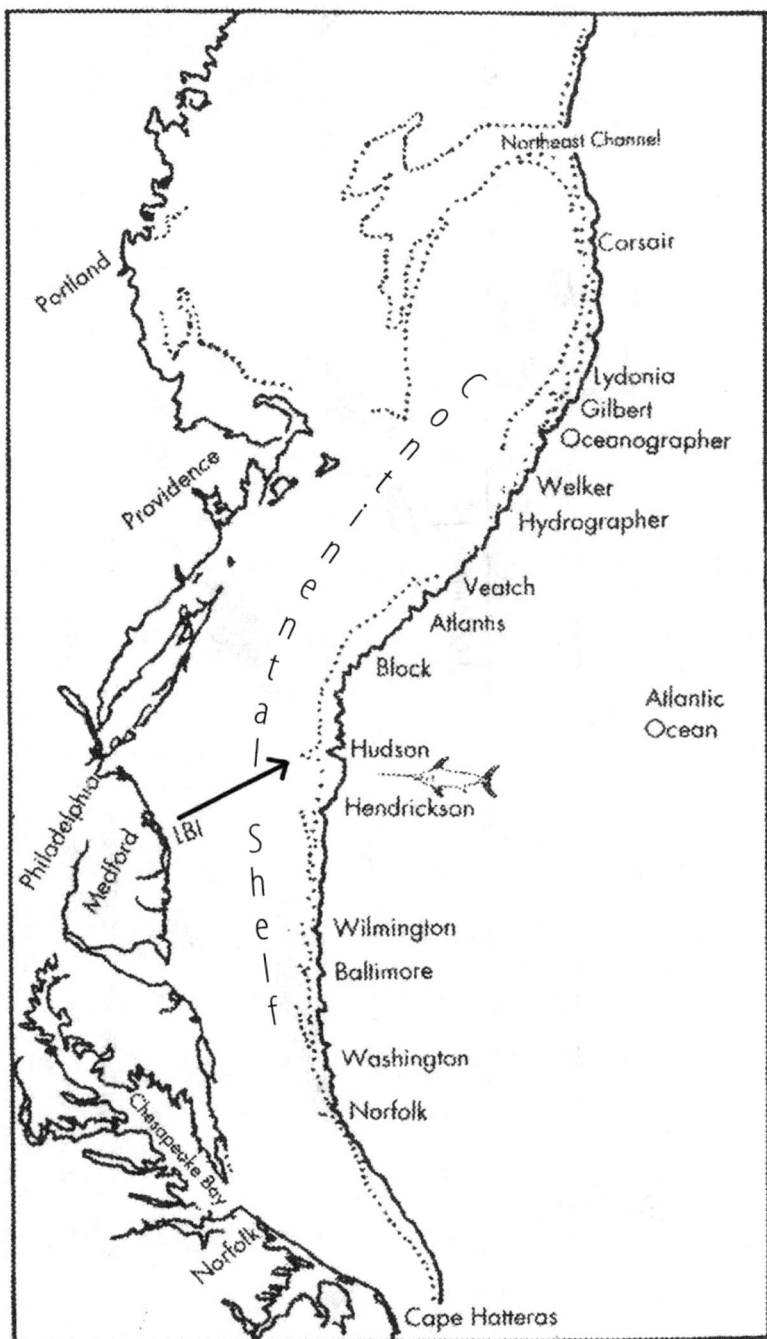

Portland

Northeast Channel

Corsair

Providence

Lydonia
Gilbert
Oceanographer

Welker
Hydrographer

Veatch
Atlantis

Block

Atlantic
Ocean

Continental Shelf

Philadelphia

Medford

LBI

Hudson

Hendrickson

Wilmington

Baltimore

Washington

Norfolk

Chesapeake Bay

Norfolk

Cape Hatteras

Long Line Set

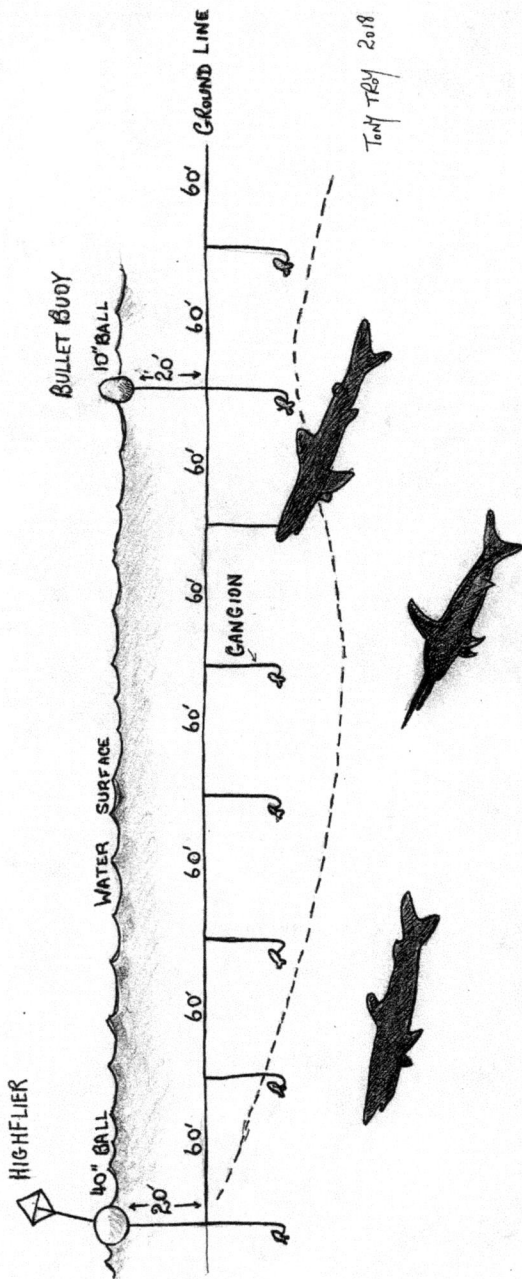

HIGHFLIER
40" BALL
20'
WATER SURFACE
60'
60'
60'
60'
GANGION
60'
60'
BULLET BUOY
10" BALL
20'
60'
GROUND LINE

Tony Troy 2018

A typical long line set tends to increase in depth the longer the gear is in the water. The catenary can increase dramatically from the weight of the fish on any given section.

West Coast Gillnet Fishery - An Example of By-Catch

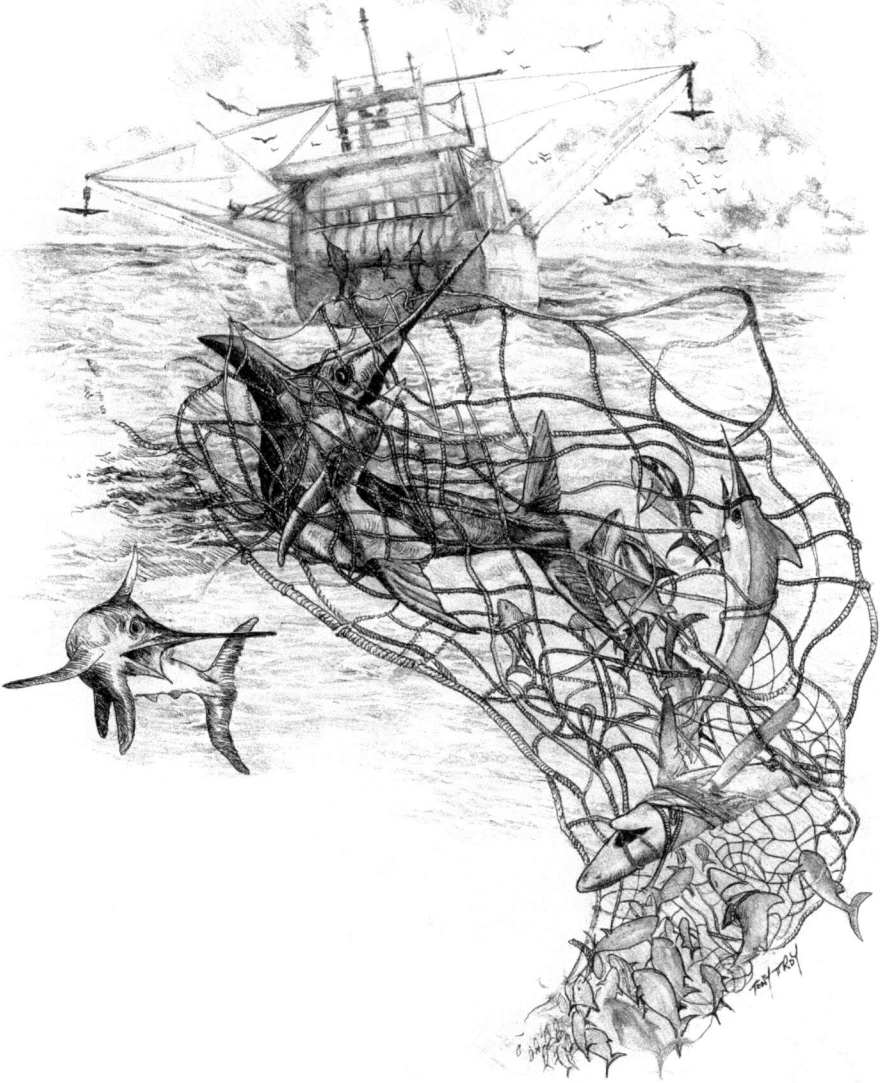

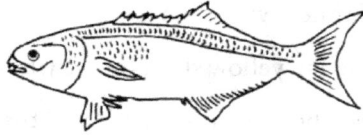

CHAPTER 4
Piranha of the Sea

Proper introductions complete, the captain puffed on his cigarette and started the diesels, both belching smoke as we cast off our moorings. I tried to look knowledgeable, tugging on lines and checking that others were appropriately coiled and free of seaweed or other debris. Various longline crews waved from adjacent boats, still tied at their berths and preparing for their own departures. Several party boats packed with summer patrons swilling their breakfast Budweiser called or waved, with what seemed respect and admiration.

As dark rain clouds gathered to the east, and curious gulls circled the vessel, we checked our progress against Old Barney's silhouette, then straightened our course toward Barnegat Inlet, maintaining a no-wake speed at five knots. Reaching the inlet we began to ease our way through the myriad of weekend fluke fishermen who seemed to own the channel in their small fiberglass-hulled motorboats. With as many as six rods in the water, these saltwater fishermen were reticent to move aside, forcing the *Defiance* to weave her way through the traffic. One fisherman in particular seemed perturbed at our presence. He was 'clever' enough to have anchored in the center of the channel, where the deeper water allowed larger boats to clear the inlet. Dressed in yellow shorts and a white stained tank top, he offered several expletives before pulling anchor at the sound of the air horn from our bridge. With an unkind gesture and lasting stare from the fluker, we eased past the vulnerable small vessels and began to feel the power of the waves as the awesome energy of the open sea pounded the steel bow of the *Defiance*.

The ship began to rock, and the deafening sounds of working engines groaned in protest. The force of the great Atlantic Ocean now pressed in on our bow as we met wave after wave with a shudder.[5] The comfort afforded by the harbor evaporated as three-foot-seas and fifteen knot winds smashed into our faces, further proof that my life of warmth and safety ashore was on hold. As we passed between the rocks, two enthusiastic jetty fishermen stared in disbelief at Jack as he heaved his breakfast into Barnegat Inlet.

Steering a northeast course, we left Old Barney in the distance. I poured a steaming cup of coffee for the captain and carefully balanced the brew as I headed for the wheelhouse, a simple act that provided a newfound respect for flight attendants. I found Captain Coyle chatting over the VHF radio. He nodded when I offered the coffee. His conversation sounded casual and

---

[5] Second only in size to the mighty Pacific, the Atlantic Ocean has a rich history of storms, shipwrecks, and maritime disasters. Composed of more than thirty-one million square miles, depths up to several thousand feet, and seawater temperatures averaging in the fifties, the Atlantic is a very inhospitable place for a warm-blooded biped with lungs.

amiable. I overheard "tilefish," which didn't make sense. We were fishing for the great broadbill swordfish, the master of the predatory fishes, gladiator of the seas, so the captain's interest in a lowly tilefish seemed perplexing. As I edged closer, he drew out a chart and unfurled it on the table. He hurriedly marked the numbers heard over the radio onto the chart, then adjusted the LORAN, and our course, due east.

A vast assemblage of deep water trenches, similar in appearance to the Grand Canyon, sits at the end of the continental shelf of the Atlantic Ocean. The inshore waters of the shelf are relatively shallow, extending to about six hundred feet of depth, where various prey fish abound. This is the location of the Hudson Canyon, approximately eighty miles off the coast of New Jersey, and this, marked in bold on the captain's chart, is where we were headed.[6]

———⬥⬥⬥———

After the captain completed his radio transmission, he turned and plotted our targeted destination into the LORAN, at the same time pointing the bow at twenty-six degrees northeast. He then requested that I summon the crew to the wheelhouse where we proceeded with our first captain's meeting of the trip.

---

[6] Farther out, however, the dark offshore canyon waters of the Abyssal Plains are several miles in depth and relatively unexplored by man. Many believe there are numerous monsters in these depths as yet undiscovered.

His plan was summed up in a simple phrase: "The tiles are hot, and the swords are not." We would fish for tilefish at dawn. I was disillusioned and wanted to dispute the proposal, but I knew my position was tenuous at best. The crew noted my disappointment, but remained silent as we disbanded to our individual tasks. There was little to occupy me at this early juncture of the voyage. The captain had a boat to run and Cooters was busy assembling rigs. Both declined my offer of help, only requesting that I "not get into trouble." It was abundantly clear that no one felt comfortable with a greenhorn on board; my job was to stay out of the way.

I struggled to the stern, still finding my balance on a relatively placid ocean, to find Jack standing along the rail actively retching into the wind. I felt the mist of digested stomach contents on my skin before I could move clear. I knew his mental anguish of missing his family, was only exceeded by intense seasickness. The common suggestion of watching the horizon didn't seem to help, and earlier ingested seasick pills proved worthless. With a final heave, he nodded a weak smile and staggered back to his rack.

I searched the boat for activities while the others remained occupied. I had packed numerous books, including *Peterson's Field Guide to Atlantic Coast Fishes*, *The Cruel Sea*, *Moby Dick*, and my trusty logbook. With no one to talk to and no interest in reading, I soon found my afternoon's amusement. The engine room revealed a treasure: a trolling rod with a Penn 30 gamefish reel. I searched for lures and found two dozen, lying tangled and rusted in a white bucket. I chose a beautiful, foot-long, aqua-green-and-white squid rigged with a clean 6/0 hook concealed in the tentacles. The captain, peering down from the bridge, warned me that at a cruising speed of ten knots we were moving too fast for effective trolling. He also explained that our purpose was commercial fishing and stressed the point that "we're not on a God-damned pleasure cruise." If I did manage to hook a fish, he continued, he would not stop the boat and lose time to land "a lowly bluefish." As I turned to head for the rail, he mumbled, "Be sure the lure is just below the crest of the wave."

Tying the squid lure to the thick monofilament line, I used my improved clinch knot technique, a time-proven method of securing the hook to the line with maximum fastening power. Double-checking the terminal tackle and testing the hook for sharpness, I was reassured by a pinpoint of blood on my fingertip. Rushing to the stern, I nearly knocked Cooters overboard. He laughed and reiterated my dilemma if I were to hook a fish, explaining that the force of a moving boat, combined with the pull of a fish, would create tremendous drag on the rod. This force, he warned, would make it nearly impossible to pull any size fish to the boat. He suggested I read a book. I nodded in enthusiastic agreement—then hurried to drop my line.

With dreams of trolling for giant marlin, I tossed the shimmering green-and-white squid into the wake and watched the lure disappear into the dark waters. Releasing the reel into free spool and protecting it against the infamous backlash known as a "bird's nest," or ensnarled fishing line, I allowed it to trail one hundred feet behind the boat. I locked the reel and felt the line tighten, the rod tip bending at a sharp angle from the pressure. The lure, with its mesmerizing tentacles and streamlined body arching through the waves, looked uncannily like a living, swimming squid. I imagined how this decoy would present to a billfish, the pulsating drone of the engines attracting it to the surface. Predators respond to changes in their environment that signal feeding opportunity, and the unusual drone of boat engines is no exception. Finally, after two hours of pure monotony and catching nothing, I searched for some goldfish snacks and a change of scenery.

An easterly squall loomed over dark waves and the temperature dipped noticeably. Just after 2:00 p.m., with rain blowing across a windswept sea, we entered the North Atlantic shipping lanes. Several boats had appeared on the horizon throughout the day, but now we neared the path of an oil supertanker off to the east. From the distance of half a mile, the huge tanker, one thousand feet in length carrying storage tanks seventy feet high, dwarfed the *Defiance*. I dashed for the wheelhouse and grabbed the greasy black binoculars from atop the chart table. Looking due east, I watched as the behemoth drew into view and could see several crewmen standing outside the bridge, casually observing us. I was startled to see the face of a crewman looking directly at me through his binoculars. I waved, but he remained motionless, then quickly departed inside the bridge. As the ships passed one another, the wave caused by the tanker hit the *Defiance* broadside, creating a heavy roll to portside. Several minutes later, we watched the tanker outdistance our vessel.

◆———◖⊗⊗⊗◗———◆

The sheer size and weight of an oil tanker are a tribute to man's ability to build larger machines, and to our increasing technology. Modern tankers can carry two million barrels of oil and transport much of the oil used in the United States. The invisible cost to the environment, however, is astronomical in scope. Prior to 1990, the majority of Americans who spent their leisure time at the beach, fishing or swimming, were unaware of the potential dangers of oil pollution from tanker accidents. Many oil spills had occurred in foreign seas, and most Americans dismissed this as someone else's problem. But with the 1988 Exxon Valdez disaster in Alaskan waters, the threat was now in our backyard, and it created a public outcry.

Eleven million gallons of Alaskan crude oil leaked from the ruptured hull of the ship and, within two months, was driven along a path stretching 470 miles southwest. Although the initial cleanup of the spill took three years, at a cost of over $2.1 billion, and the impact on the area's wildlife and human population was tremendous, the terrible accident brought the issue to the attention of a previously uninformed public.

The Exxon Valdez spill led to new government regulations designed to prevent this type of future disaster. The most widely recognized of these new laws was the Oil Pollution Act of 1990 (OPA), which established regulations that focused on the two main reasons the Valdez disaster occurred. The first OPA provision required that all oil-carrying vessels operating in U.S. waters be equipped with double hulls by the year 2015. The Valdez was a single-hulled vessel and was easily torn when she ran aground on heavy rocks. The second reason for the accident appeared to be a fatigued crew; hence, the second provision attempted to reduce human error by limiting the number of hours a crewman could work per day. (This seems indulgent, yet the current recommendation remains an arduous twelve-hour day.)

Other solutions to prevent future oil spills currently are being tested. One area of research involves the technology to design safer tankers. Several designs show potential, including the intermediate deck design, which divides the ship into an upper and lower compartment. Although oil is stored in both compartments, a potential grounding would only affect the lower compartment, and gravity would force water inward rather than oil to seaward. Also in the works is the possibility of under-loading tankers to keep the oil level below the waterline. Naturally, this concept has proven very unpopular with the oil industry, as less oil carried means less profit. Further studies by the Navy, the Coast Guard, and oil conglomerates should lead to future improvements, although the best defense is avoiding accidents altogether by decreasing human error.

The risk of another catastrophic oil spill is omnipresent within the United States. It is less a matter of *if* and more a matter of *when*. Decades later, the environment in the area of the Exxon Valdez disaster has failed to show complete recovery and, in spite of the five billion dollars in damages that the federal government ordered Exxon to pay Alaska, both physical and economic scars remain. How many of us can forget the images of millions of dead and dying seabirds, trapped and struggling in an oily bath; whales and dolphins surfacing to breathe in the midst of burning oil slicks; and massive Leatherbacks, the sea turtles shells covered in flames?

Human consequences were also severe. Families were torn apart by loss of income, divorce, illness, and suicide, while entire towns and villages vanished or dispersed. The involvement of concerned citizens is vitally necessary to help prevent further oil disasters from occurring.

———◦———⟨⟨⊗⊗⟩⟩———◦———

The great tanker moved into the squall and soon vanished from view, leaving the *Defiance* alone on the swelling sea. Under poor visibility and mounting fog, we steamed through the shipping lanes at reduced speed, listening to the occasional foghorn of a passing freighter seen only on the radar screen as large, bright green blips. The captain spoke of the dangers of unseen vessels and talked about the consequences when ships collide during bad weather. He mentioned for emphasis the tragic tale of the *Andrea Doria*, a passenger ship that went down off Nantucket in 1956 in very heavy fog. She had crossed the Atlantic exactly one hundred times and was thought to be virtually unsinkable, until the heavily reinforced icebreaker bow of the Swedish liner *Stockholm* rammed her broadside, causing her to capsize and eventually sink. Now mentioned by historians with the *Titanic* and the *Lusitania*, the *Andrea Doria* was the worst passenger ship collision in history. From 1919 to her sinking in 1956, twenty-seven million people had been transported by passenger ship with not a single death or injury caused by collision. Yet, fifty-one people died as a result of this one.

Due to the poor weather conditions the captain assigned me to the upper rail as lookout. Donning heavy rain gear, I moved to the bow to scan the horizon for traffic. Under light rain and a cold, salty wind I could pick up the sound of waves and an occasional squawking gull over the engine's rumble. But when I glanced toward the stern, I noticed the tip of my long-forgotten fishing rod bending violently. The squid was no longer visible at the surface. I knew I had hooked my great fish. Scrambling to grab the rod, I jerked it from the holder and was almost pulled into the sea under the heavy strain of both fish and boat. Searching in vain for a rod belt, I debated cutting the line and surrendering my prize. I strained and continued the battle for a full ten minutes until I heard Cooters shout, "He's got a fish on." I was red in the face and nearing the breaking point when the captain yelled over the PA system: "Say uncle!"

Suddenly, the boat slowed to a drift as the engines dropped into idle. Imagining a giant mako shark, swordfish, or tuna, I began to recover my line. Cooters joined me at the stern, lighting a cigarette with casual indifference. He only said one word: "Bluefish." But there was no way such a titanic struggle could be caused by a small bluefish. I weighed 160 pounds, so I knew this had to be much larger than a blue. But he was right. The rain-swept water soon revealed the sleek, steel gray shape of a slammer blue, weighing about ten pounds. The fish finally surrendered at the surface, exhausted after a ten-minute struggle[7].

---

[7] The bluefish (*Pomatomus saltatrix*), called the piranha of the sea, is regarded as one of the world's greatest game fish. Reaching a size of over thirty pounds, these sleek torpedoes are found in scattered distributions worldwide and in both temperate and tropical oceans. They group in massive schools and attack nearly anything they can catch, mutilating prey by pure instinct. These predators are so aggressive that they have even been known to attack bathers on occasion. Saltwater fishermen respect the bluefish's tenacity for life on hook and line. Although oily and not very palatable, they are often killed for the mere sport. Unfortunately, many bluefish will be discarded in garbage landfills, a tragic waste that could be prevented by responsible fishing captains.

To my great surprise, Cooters exhibited tremendous respect for this marine predator. In the driving rain he slowly, gently, hand-lined my catch to the deck, protecting his hands with an old pair of torn leather gloves. My initial jubilation turned to despair, knowing that my epic battle had been fought over a mere ten-pound fish. Once on deck, the fish writhed, blood streaming from the corner of his mouth. Cooters carefully removed the hook with steel pliers, prudently avoiding the razor-sharp teeth. The fish immediately showed its displeasure by disgorging its lunch on my white sneakers.

I wanted it released and Cooters obliged by dropping the writhing fish headfirst into the churning sea, where it disappeared with a single flick of a forked tail. Cooters mumbled a conciliatory, "He'll live," and marched off to resume his work. On my damp shoes, I noticed nearly a cupful of fish guts, including herring chunks and small white pieces of squid, evidence that this fish deserved its reputation for gluttony.

My fishing ended, I pulled in the line, rinsed off the reel and my sole pair of sneakers with the freshwater hose, and stored the gear in the engine room. Not wanting to further aggravate the captain, I made a mental promise to desist from further sportfishing for the remainder of this trip. We had resumed our course and speed as the rains lightened. The sea surface was now empty. Well beyond the sight of land, there was no visual stimulation to provoke interest. I searched the horizon for the blow spout of a whale, the crash of tuna chasing surface bait, or the fin of a shark. Not an animal in sight.

These fish were once believed to be an inexhaustible resource, but now have been so heavily exploited that numbers are on the decline.

Now close to 4:00 p.m., I went up to the wheelhouse, hoping to talk with the captain. Instead of fishing, he discussed food and questioned my cooking skills. Although not experienced beyond the comfort of a cheeseburger, I exaggerated my previous culinary masterpieces; he promptly assigned me dinner detail. In his best pseudo-British accent, he informed me that we "dine at six" and warned of the incoming low-pressure front expected, otherwise known as bad weather, for early evening. As I grabbed the ladder, he suggested fried chicken, broccoli, and mashed potatoes.

By 5:00 p.m., I scrambled to complete the dinner assignment just before the storm intensified. Bright white streaks of lightening ripped across the dark skies. I counted the thunder as only five miles distant. Within minutes, the rain began to soak the boat as she heaved and rolled in the churning water.

At 6:15 p.m. I announced "Food's ready!" It had a familiar ring of home. I was surprised to feel elated. Here I was, in the middle of the Atlantic, working on a fishing boat with three strangers, but this was my first contribution to the crew all day and I was proud of the results. Unknown to all, this was the first real meal I'd ever cooked, and it actually appeared reasonably appetizing. Cooters was quick to the table and requested that I take food to the captain, who was unable to leave the wheelhouse. Carrying a plateful of hot food in one hand and grasping the ladder with the other, I performed a balancing act that required me to lean with the ship. As a result, the plate made its way to the captain with only half a cargo.

Satisfied that the captain had what he needed and was settled at the helm, I continued downside to share dinner with the crew. Only Cooters was present. Jack remained curled up and seasick in his bunk, refusing both food and conversation. But Cooters hungrily devoured everything in sight, relishing my culinary efforts. He openly discussed his personal history and life at sea and as we became better acquainted, he seemed to appreciate my marine science background. I, in turn, noted his remarkable knowledge of, and respect for, the fish we hunted.

Cooters began cleanup while I took the captain a cup of hot coffee and dessert of fresh strawberries with whipped cream. Fresh fruit was a special treat on board. Depending on the variety, fruit typically lasts for about a week and cannot be preserved much longer due to lack of space. The demand is high, but the captain always receives first dibs. Finishing the berries, he signaled approval with a belch and a grin.

The storm abated during the next few hours, allowing the crew to settle in. The captain and I talked in the wheelhouse over the sweet sound of Jimmy Buffett until almost 11:00 p.m., when he warned me to "hit the rack." We were nearing our destination and would start fishing at sunrise. As I climbed into the top bunk, I noticed that my rack mate was out cold, still dressed in his jeans and sweatshirt. A yellow, stained puke bucket was tied to the post next to his head. I felt a twinge of empathy for him only moments before I fell into a deep sleep. With calming seas and dawn only hours away, our first day of tile fishing was nearly at hand.

CHAPTER 5
The Tilefish Blues

A sharp knock startled me out of a fitful sleep.

"It's 5:30 ladies—out of the rack. We got work to do."

It was pitch dark. The captain disappeared through the galley. I felt annoyingly tired, but the pungent smell of coffee brought me to my feet. My bunkmate was rousing and appeared to be in the same sleep-deprived stupor as myself. The captain and Cooters had shared most of the evening watch, so although I felt tired and irritable, I realized that I had no reason to grouse. The captain had been up all night, Cooters had barely two hours of sleep, and Jack was still sick as a dog.

Managing to hold a cup of piping-hot-black coffee, I struggled up the ladder. The ship was slowly circling a designated area, and the captain was checking both ocean temperature and depth. We were in six hundred feet of water, at the edge of the continental shelf, now ninety-eight miles off the Jersey coastline. The seas had calmed to four feet, and a gentle breeze blew northwest at six knots. The VHF radio blurted the NOAA marine forecast as "favorable." As the sky brightened slightly, I could see Cooters moving cardboard crates of mackerel to the stern bait table. Jack stood nearby, quiet but seemingly ready to work, wearing orange oilskin suspender overalls and black boots. He apparently was over the worst of his nausea. I donned oilskins and boots, sucked down the last of the lukewarm coffee, and raced back to my bunk, searching for a particular book I had brought aboard. Looking for the section on tilefish, I thumbed through my copy of *Peterson's Field Guide to Atlantic Coast Fishes*. Then I heard it.

"Attention, attention. All greenhorns on deck."

I immediately tossed the book and scrambled topside to begin my first day of work at sea, wishing I could have read something about the prey before setting off to kill it.[8]

---

[8] The tilefish (*Lopholatilus chamaeleonticeps*) lives in the deep waters off the continental shelf from Nova Scotia to Florida, in water pressure that is twenty-seven times that of the surface. They can attain a size of three feet and a weight of fifty pounds. A massive die-off in 1882 created a newfound awareness of the dense populations of this untapped deepwater species. Presumably, an ocean anomaly of that year created an unusually cold deep current, causing the massive kill. Ships' logs from that era tell of miles of dead and dying animals and the resulting stench. Today, tilefish numbers, like most commercially valuable food fish, have plummeted due to overfishing (though not abundant, tilefish are still pursued by longliners due to their high value as a delectable fish). Since a key to the health of a species stock is the average size of captured fish, and the typical tilefish caught today weighs only about four pounds, we know that this weight and size relates to a reproductively immature animal. The species has been gravely exploited, and the stock may never fully recover.

Joining Cooters and Jack on the stern, we began the process of chopping bait. Jack dumped a fifty-pound box of frozen mackerel on the bait table and we cut through each to make five steaks per fish. These mackerel sections then were placed on circular-shaped hooks the length of a finger. Monofilament line ran from the hook to a snap, which eventually would be clipped to the outgoing ground line uncoiling from the drum. As soon as the captain gave the "okay to start" call, Cooters dropped the first hiflyer into the sea. It fell into the wake like a huge bobbing cork. He then tossed us bags of rubber bands, instructing us to band our oilskins to our boots to prevent waterlogged feet.

The engines hummed in slow gear as we set hook after hook, dropping each into the deep water. A thirty-pound lead weight was snapped to the ground cable overboard to force the line to the muddy bottom, some six hundred feet beneath our keel. Attaching snaps with hooks at twenty-foot intervals, Cooters continued the tedious task while Jack and I continued to bait hooks at a frenetic pace.[9]

---

[9] Setting gear is one of the most dangerous jobs onboard and requires a great deal of caution. Many men have been injured or killed by getting snagged with a sea-going hook and falling overboard. If the unfortunate fisherman should get deeply hooked and fails to cut the line, he could be pulled to the bottom. This is a horrible way to die, given the long moments to reflect on the way down. Another danger is with the oilskins themselves. Although oilskins function to keep a surface dry, they also fill with seawater once a man has fallen overboard, acting as an anchor. Even if the fisherman is able to cut the line before going deep, he may drown before the boat has time to recover him. Sebastian Junger's "The Perfect Storm" graphically illustrates the experience of a commercial fisherman falling overboard.

The monotonous process continued for the next three hours. The only break in the routine was an occasional hiflyer overboard, attached to the ground cable and used to mark our line. I had expected this trip to be a nonstop thrill ride, but quickly learned there would be great periods of intense boredom followed by moments of pure adrenaline. After three hours and twenty miles of longline, the last hiflyer went into the sea, and the *Defiance* separated from the line. The longline now was floating independently of the ship, and the ability to maintain visual connection with the line became imperative.

Jack went inside to make lunch while we hosed down the deck, stored loose gear, and prepared the boat for the eventual haul. The lunch table bulged with stacks of bologna, American cheese, corned beef and pastrami, bags of greasy potato chips, dishes of coleslaw, potato salad, and fruit, with a milk jug as the centerpiece. Jack looked somewhat improved. He finally seemed ready to talk, and I found his story rather pitiful. He lived in New Jersey, was thirty-five years old, married for ten years with two children, and worked in construction. But he never felt very happy with his occupation. Although he had absolutely no sea experience, he always had been fascinated with the ocean. He avidly watched National Geographic shark specials and read sea novels. Jack's father loved the sea and disapproved of his son's career choice away from the water. When I boldly asked why he elected to change his life so drastically to work on the sea, he simply replied, "It was time for a change." It seemed an odd decision with a wife and two children to support, and I suspected he was doing this more for his father's approval than for any other reason. Despite our differences in age, education, and family responsibilities, it appeared that we were not so dissimilar. It was our fathers who had unwittingly shaped our decisions for this shipboard experience.

The five-hour soak was over and we returned to the deck to prepare for the haul. It was time to retrieve the gear. The captain climbed to the wheelhouse, the engines roared to life, and we steamed the half-mile to the lead hiflyer. Six hundred feet below, struggling fish, impaled on steel hooks, fought for freedom. Donning leather gloves, we prepared for the start of a grueling four-hour pull.[10]

As we glided up to the first hiflyer, Cooters adroitly snared the gear with a ten-foot gaff pole. He quickly pulled the flyer to the stern through a door located on the port side, unhooked it, then handed it to Jack, who ran it to the bow to be secured. The captain, operating the boat from the deck, used a hydraulic winch to retrieve the line. He had full control of the boat and winch, and a full view of the action. Using a lever, he could speed or slow the incoming line and observe the crew at the same time. The ground line wound back on the main spool as the first snap became visible at the surface. He tugged on the line to determine the presence of a fish, a taut line signifying a catch.

As the first snap presented at the surface, the captain tugged, then simply said, "No fish." The bait appeared stripped from the hook, and Cooters

---

[10] The up-haul, or gear retrieval, is regarded as a labor-intensive stretch of fishing, taking nearly twice as long as the downhaul. Miles of line are retrieved slowly, at about five miles an hour; extra time is required to secure the fish in the boat and re-spool the gear in an organized fashion.

wound the line back onto the reel. The routine always was the same: The captain retrieved the line with the lever, checked for the weight of a fish, and handed each successive empty hook to Cooters.

The crew was intensely quiet until the sixth hook. The line appeared taut and, with a tug, the captain announced, "Fish on." We had our first fish of the day! (According to the owner of the *Defiance*, for whom we all worked, fish meant money, and big fish meant more money. Although not particularly concerned with money, I was elated at the opportunity to work with my first commercial catch). Captain Coyle looked at me. "Start gutting the fish," he ordered. I reached for a foot-long fillet knife and, sitting on a milk crate, prepared to receive the flopping, struggling life-form.

The first tilefish landed unceremoniously on the deck with a dull thud. It weighed fifteen pounds and displayed brilliant yellow and green. The swim bladder, used by the fish to regulate buoyancy, grotesquely protruded from the mouth like a white balloon. It had been down at a depth of six hundred feet before being rapidly pulled to the surface. This had caused gas in the bladder to expand much the way divers get the bends if they rise too quickly, no doubt a painful process that extruded the bladder through the fish's mouth.

I again was instructed to "gut him." I hesitated. I had fished before, but this was different. I had caught bluefish, unhooked them, and put them in a cooler, allowing them to suffocate under ice, perhaps not much better, yet an improvement on being sliced alive. The instructions were repeated yet again, so I performed my duty. I placed the tip of the sharpened steel blade into the throat and slid the knife along the underside of the belly. A fatal twelve-inch incision caused the fish to jump. I grabbed the intestine, liver, and kidney and ripped them from its quivering body. A final shutter, and the fish was tossed back to the stern. I felt ashamed to have killed a living creature in such a brutal manner.

As the captain pulled another hook, he was displeased to see it was a dogfish,[11] a small shark that schools in great numbers, knowing it could represent an infestation. The hook was buried deep within the head of this unfortunate animal. The shark bled profusely as the steel was torn from the fish with heavy pliers. The two-foot shark was then tossed back into the sea, unlikely to survive the trip to the bottom. It seemed such a waste.

Finally, after five hours of pulling gear and packing, we prepared for deck cleanup. The captain was visibly aggravated with our poor results. We caught seventy-five tilefish on twenty miles of longline—an economic disaster. The meager thousand pounds of fish was insufficient to even begin to show a profit.

Cooters climbed into the hold as we handed him each fish, individually. He placed each one belly-up on the ice, lined like stacks of cordwood. The captain already had started to plan for our next set while Cooters and I continued to store the fish, and Jack fixed supper. Finally, we completed the packing, hosed down the decks, and stored the remaining lines and gear.

---

[11] Though there's no American market for dogfish, it would be relished in England as fish and chips. (As opposed to the British, such waste typifies the finicky nature of the American palate.)

During dinner, the captain claimed responsibility for our dismal catch, and discussed his plan to steam to a more productive location. He seemed very interested in a stretch of water located to the northeast, an hour away, and his enthusiasm rekindled our spirits. We reeked of tilefish, but even the dried scales on my hands didn't interfere with the consumption of my twentieth and final rib. After a final glass of milk, I started cleanup. We had worked from 5:30 a.m. to 10:00 p.m., considered a short day for a fisherman. Jack hung out as I washed dishes. After only two days at sea, he now regretted the decision to go commercial fishing, and admitted that he missed his wife and children. The reality of another three weeks at sea further upset him, yet he was stranded. Aside from a life-threatening injury, the captain would not turn back or call for a Coast Guard rescue helicopter. Jack was desperate for reassurance and we talked for another hour.

That night, Cooters and the captain shared watch while Jack and I slept. This schedule was to become a special gift; we would get little rest for the remaining three weeks at sea. I was assigned the middle watch, from midnight to 4:00 a.m. Though off for the evening, my watch would start the next day and extend thus for the duration of the trip. I was pleased to work the middle watch; it would afford me a solid hour and a half of sleep before the following workday. I settled into the rack and turned on the radio to the soothing sound of the marine forecast: "From the weather station at Barnegat Light, the offshore marine forecast is for seas of six to eight feet, visibility ten miles and winds at twelve to fifteen knots." It was an unfavorable forecast. Around midnight, I felt the *Defiance* slow to a crawl. We had arrived at our destination, and I prayed tomorrow would be a better day as I drifted off to sleep.

The knock on the rack shocked me from an unusually peaceful slumber. After a quick glance at my now scale-encrusted dive watch, I silently cursed the captain and eased my aching body toward the floor. In the rack below, Jack coughed and gagged. We were both overcome with that exhausted feeling when the desire to sleep overwhelms the need to rise. It was cold, damp, and dark outside the bunk, and the thought of another day of struggling, bloody, tilefish slapping on deck left me numb. I felt momentary regret that I had decided to spend a month at sea and painfully realized this was only day number three.

Jack and I staggered to the deck and launched into yet another brutal twenty hours of tilefishing the Northwest Atlantic. By the end of the day, we all were in a foul frame of mind. Jack was miserable from intermittent vomiting, fatigue, and loneliness for his family. I was the petulant child, pouting because we weren't swordfishing. The captain and Cooters were disappointed that we had wasted enough bait for three thousand pounds of tilefish and caught only a thousand. In addition, the weather was abysmal. The captain called for a second crew meeting over dinner.

Because of the rough weather, our dinner was simple; hot dogs, baked beans, and corn on the cob. The dejected crew was silent, the scraping sound of multiple jaws to corn beginning to grate on my already frazzled nerves. Captain Coyle finally spoke. He looked at me and managed a smile. "Well, Tommy, got great news for ya. We're steamin' east to deeper water—for swords. Are ya happy now?" Frustration lifting and immediately relieved, my long-anticipated swordfishing encounter was about to become a reality. From that moment forward I attacked every assignment with renewed vigor.

The *Defiance* steamed east into moderate seas, heading for the fabled swordfish grounds of the Hudson Canyon. Late into the night, two longliners from New Jersey joined us, their lights twinkling on the horizon some five miles off our stern. Our little flotilla was a spirited group, talking nonstop over the radio and all heading in the same direction, with promising waters ahead.

CHAPTER 6
In Pursuit of the Mighty Beast

Moving farther into the Hudson Canyon, the easterly course took an hour, and it was dawn when we finally reached our destination. In the meantime, the temperature had dropped twenty degrees. Although it was early July, the thermometer read only 60 degrees. We faced twenty-knot headwinds and six-foot seas under a cloudy morning sky. The plan was to prepare the first set two hours before sunset. The gear would soak overnight and be retrieved at dawn the following day. Over breakfast I devoured bacon, corned beef hash, four fried eggs, and four slices of buttered toast. After topping off the meal with two more cups of coffee, I was ready for my first day of swordfishing. I donned my heavy denim jacket and two pairs of long johns and joined Cooters on deck to prepare bait. In spite of an icy northwest wind and an approaching storm, we finally were going for the "oceanic gladiator," the Atlantic broadbill swordfish.

We worked until early afternoon, then had a couple of hours to kill before the next assignment. Jack returned to his bunk and slept, while

Cooters and I sought out the company of Captain Coyle in the wheelhouse. Believing we were on our way to better times, our moods were lighthearted and relaxed. Cooters and I struck up a decent harmony with "wasting away on Cap'n Coyle's fishing boat," to the tune of one of the captain's favorite songs by Jimmy Buffett.

Now finally positioned over the inside edge of the Hudson Canyon, the captain and I searched the charts for the most productive stretch of water. He explained that the 60-mile-long canyon cuts into the continental shelf about 120 miles southeast of Sandy Hook, New Jersey. Originally carved by the Hudson River during prehistoric times, the canyon stretches from the one-hundred-fathom (six hundred feet) chart line into the black abyss, where waters run over a mile deep. An inshore section is called the Mud Hole/Gorge, while the deeper offshore is known as the Hudson Canyon.

Here, where temperature gradients are vast, zooplankton, or microscopic animal life, migrate to the surface at sunset in a process known as vertical migration. Small baitfish come to feed, followed by larger predators such as mackerel, squid, and herring. These, in turn, attract giant swordfish. "They be big feesh," he said with a smile in his best pseudo-Mexican accent. That long-winded explanation clarified why we now were one hundred miles off the eastern American coastline in one thousand feet of water. But we still had to find the captain's favorite stretch of water, somewhere along the canyon's edge.

Dr. Thomas Armbruster

It was 5:00 p.m. when we finally reached prime sword-hunting grounds. We talked with the captain as he pulled a dozen nautical charts from the drawer, searching for Bathy 31, his best-loved bathymetric chart of the Hudson and neighboring Tom's Canyons. Pointing to the multicolored chart at an area known as the West Wall, he described our evening set, using a black marker to represent our line, and circled a huge yellow area on the map that extended twenty miles, from the tip of the Hudson Canyon to an area known as The Letters. Our captain had found an ideal stretch of water, an area of interface where cold Labrador currents meet with warm, seventy-degree Gulf Stream waters.[12]

Peering over the map, I inquired about an area that looked like a thin, pink finger extending to the east, a channel known as The Deep, which drops into fifteen hundred feet of ocean. Depending on the current, our lines might drift across this channel overnight, placing us above nearly half a mile of cold seawater. Our fishing companions to the west were now beyond sight, meaning that the *Defiance* was so far offshore that we were completely alone, far from the warm, sandy beaches surrounding Old Barney. And we

---

[12] The western Atlantic population of swordfish relies on two massive currents (found by fishermen in the 1960s on satellite photos), the Labrador Current and the Gulf Stream. Because of their opposing flows and significant temperature differences, these waters have become an optimal feeding ground for millions of fish. An inshore current, the Labrador travels north to south, bringing with it nutrient-rich cold water. Traveling just off of Labrador, the Gulf Stream runs south to north, adding warmer water to the mix. Thus, the major fishing grounds for the U.S. Atlantic swordboat fleet are found at the eastern edge of the continental shelf, off the coast of Florida and as far north as the Grand Banks of Newfoundland.

were fishing in 250 fathoms of ocean, hunting giants. It dawned on me that we could be replaying a scene from *Jaws*. The captain reviewed faxed satellite infrared photos denoting thermal gradients or temperature differences which attract broadbill swordfish. His decision? Let's fish!

Standing at the bait table in boots, oilskins, a padded denim jacket, and gloves, I waited for the first hiflyer to be dropped off the stern and the captain to ease the bow into the swelling seas. The main drum was spinning off yards of line per second; my assignment was to bait hooks, alternating mackerel with squid. We attached light sticks with rubber bands to every other line. Known as Cyalume lights, they are similar to the glow sticks found at every country fair. As each light stick hit the water, I watched the eerie, phosphorescent green glow disappear into the dark black sea. Although the green color attracts swords, the downside to the environment is that they take nearly one thousand years to biodegrade. A neophyte commercial fisherman, I wanted to capture swords, but the environmentalist in me dreaded polluting the oceans.

I also was aware that a freshly-hooked swordfish has little chance of survival. Deprived of oxygen, the fish finally will die on the hook. Under current regulations, if it weighs more than forty-one pounds, the animal is boated for market; if less than forty-one pounds, it's released. However, the combination of trauma, blood loss, and predation affords this released fish a fifty percent chance of survival, an atrocious waste of life and a consequence of longlining.

The first nine lines went out without incident. The tenth bait was met with a sudden snap to the ground line. Cooters explained that a large fish probably had struck one of the outgoing hooks. "We'll meet her in the morning," he happily informed me. The hooks now stretched across five miles of ocean, and the baiting process continued for the next three hours, until the captain announced over the PA system, "End of set." The last hiflyer was attached and tossed overboard into the churning black sea. We packed the extra bait back into the hold and hosed down the deck. After a check of the horizon and a brief moment to admire the sunset, I delivered a cup of coffee to the captain, then headed back to the galley. It was dinner time, and my job to fix the landlubber's special of London broil steaks with baked potatoes.

———⟨⟨⟩⟩———

Drifting independent of the *Defiance*, the set clearly was visible as the last remnants of light shadowed the southernmost hiflyer. A line of stars appeared through scant clouds, and the white strobe light from the last beacon buoy softly illuminated the quiet waters. Dinner was an enthusiastic affair, everyone participating in lively conversation. Even Jack seemed more relaxed and less dejected, eating heartily for the first time during the trip. Although ever present, Mutt hungrily begged scraps and now showed little predilection toward nipping. The crew discussed our assignments for the evening, and I mercifully was granted time off until my middle watch duty at midnight.

But first, the captain summoned Jack and me to the wheelhouse to explain the duties of a watch. Since Jack was given the first night watch, from 8:00 p.m. until midnight, he was officially on duty, so he paid particularly close attention. This tutorial was only a formality, however, in that the captain always was on watch, even when observing discreetly from afar. Such behavior was reassuring, and another example of his paternal role. But when forced to intervene, he did so with much respect. For example, he'd step into the wheelhouse with an excuse, "I had to recheck our position," rather than with a more direct order intended to circumvent a greenhorn's efforts.

Evening watches, he told us, involved a series of significant responsibilities, the foremost being to visually observe the horizon for ship traffic and surface vessels that might present a potential collision. We also performed frequent radar checks, looking for bright green marks on the radar screen that represented surface objects. Maritime history is full of stories of inattentive crews being the most common cause of offshore collisions. The captain explained that we were to wake him in case of confusion and then went on to discuss the importance of maintaining visual contact with our gear at all times. Although the beacon buoys were attached to the set, the best prevention for gear loss was visual. The boat drifts independently of the set, and it requires frequent engine adjustments to maintain constant visual contact, so he demonstrated how to push the throttles forward to keep near the mainline. He reiterated the importance of watching the set and told us we would pay the ten thousand dollars for lost gear "if you screw up."

Jack took over the captain's chair, apparently relishing his new responsibility. The captain wished us good luck and left me with Jack to learn our new assignments. We reassured each other that we were equal to the task. The weather was good, the skies were light, and a gentle wind blew the seas into crests of three to four feet. I checked our numbers on the depth gauge and noticed the readings skyrocket as we drifted over fifteen hundred feet of water, all the while maintaining visual contact with the white, intermittent flashing of our gear. We described the giant fish that might be hitting our hooks and listened to a favorable forecast for the next day's pull, and we talked about home. In the background, the marine band radio provided weather and shipping information as well as entertainment, and stories of cheating wives, drunken crews, and newfound money filled the dark, cold wheelhouse. After an hour, and with a "Good luck," I left Jack to his sea duties and went for two hours of sleep before my own watch would begin.

Just as I nodded off, I felt a tap on my shoulder. A quick check of the time revealed that it was midnight, and time for my watch. Fatigued and dull with sleep, I pulled on my scale-encrusted Levi's, donned a sweatshirt, and headed topside. My first obstacle to a successful watch was not lack of experience or exhaustion but Cooter's mongrel dog, Mutt. He blocked my path, growling, his fangs barred. I was tired, achy, and much of my good clothing was shredded, thanks to this beast. Unprepared for battle, I tossed a few scraps from my pocket, distracting him long enough to gain the ladder to the wheelhouse.

Only the green glow from the radar and a small bulb on the chart table illuminated the wheelhouse. The *Defiance* rocked gently in the swells with good visibility, making it easy to locate our set. Several dim lights to our northeast indicated we had company. It was difficult to determine their identity, but a quick radar check indicated eight miles distance between us. I searched for them over the radio but heard only silence. Scanning over black waters, the white blinking beacon buoy reassured me, and I was psyched to be entrusted with such heavy responsibility.

As the tedious watch hours passed, my mind began to wander. Living at sea was much more exhausting than I'd ever imagined. The constant motion of the boat demands balance and saps energy. The temperature offshore feels colder because of the moisture in the air, and the chill tends to draw energy from the body. In addition, the strong sunlight reflected from the water burns the eyes, and the constant drone of engines from hours of steaming causes mental fatigue. Even at a dead stop, when the engines are off, the ship's generator runs constantly, emitting an incessant, nauseating hum. The combination of mental boredom, loneliness, and the physical challenge of long hours of hard labor all combine to cause tremendous fatigue at sea.

I wondered how the folks at home were doing. I reflected on my life. I worried about my future and tried to imagine where I'd be working after graduation. Perhaps I would get my doctorate at Woods Hole, get a professorship and teach in New England. I daydreamed of summers spent in a submersible while studying the giant squid, scuba diving, and long workdays on the beach. My dreams of getting paid for something I loved had become obsessions over recent months.

But a recurring, nagging voice stated, "Your head is filled with seawater." It was the ever-present voice of practicality issued by my strict German father. He often stated that an education in marine biology would do nothing but "get you a job on a shrimp fishing boat." It was a constant struggle to decide to chase my dreams or settle for a good paycheck in a dull job. Commercial fishing doesn't necessarily offer a steady income, as profits can be several thousand dollars or merely room and board for a month at sea, depending on the quantity and quality of the catch. Oftentimes, the pay is the shirt off your back when times are lean.

I had discussed the issue of pay with the captain prior to departure. It seems that the most junior deckhand, without a family to support, would receive the joker's ration of the spoils. This translated into eight percent of the *profit*. Considering the cost of fuel, insurance, bait, ice, gear, and groceries, especially groceries, we were currently many thousands of dollars in a financial hole. The expense of fifty pounds of grade 'A' steak alone was enough to flatline our tally sheet into the red. With only one thousand pounds of fish in the hold, on day three at sea, I made no immediate plans for retirement.

❈

Imagining our profits swimming under the keel, I envisioned a huge school of prime giant bluefin tuna darting toward the gear. Finally, at 3:55

a.m., the captain relieved me. I was thankful to have survived my first watch and, as I crept down the ladder, carefully scanned the dim galley for signs of Mutt. Cooters was awake, pulling his tattered, blue Viking Village sweatshirt over a heavy white T-shirt, preparing for another long day at sea. As my eyes adjusted to the dark, I noticed Jack curled into the fetal position, the stained puke bucket at his side. He was buried under several blankets, and the chill morning air forced me to do the same. I wondered about the moment in my life, the very stupid moment, when I chose the 'glamour' of commercial fishing.

It had been a twenty-hour day, and I had an hour to nap before beginning a new one. I closed my eyes but was unable to sleep. The point was moot; it soon was 5:00 a.m. and time to start again. I dragged myself to ol' Betsy, our trusty, rusted coffeemaker. She spurted to life, and her perfumed aroma soon filled the air. At this point, she was the only pleasant-smelling thing on board. I longed for warmth, and a hot shower. I so yearned, at that moment, to be home, and I ached at the thought of the lonely weeks ahead. It was only our fourth day at sea. I balanced my coffee carefully and pulled my way up the ladder. Cooters was checking our LORAN numbers and fixing our position. He smiled at the sight of the Styrofoam cup and greedily snatched it from my hand before I could utter a word.

In the brief time we'd been at sea I had become accustomed to everyone's preferences, especially for the jet fuel version of coffee we incessantly craved. This readily available source of caffeine was more important than water to the crew. We shared a liking for the strong, thick brew, heavily laden with sugar. It had a sustaining and revitalizing power both physiologically and emotionally, and I took great pleasure in serving it. It made me feel as if I was contributing to the team. The crew seemed to appreciate my efforts, and my behavior was positively reinforced, not unlike that of Pavlov's famous dogs.

———— (◊◊◊◊◊) ————

We were positioned over the Hudson Canyon, some ninety-six miles east of Barnegat Light. The cool salty breeze drifted through the wheelhouse as the faint glow of a spectacular eastern sunrise filled the windows. Under a sky of fading stars and a half moon, the eastern sky brightened with streaks of orange and red over low clouds, reflecting off the surface and limiting visibility. The sunrise was beautiful, even romantic, but I had to share it with the unshaven, grease-covered, pungent Cooters, not exactly my first choice for romantic companionship. I voiced my thoughts, drawing a bright smile and hearty laugh from him as he nodded in agreement.

The *Defiance* floated one-quarter mile from our gear. Using binoculars, I followed the serpentine line well into the horizon, noticing intact hiflyers and bobbing corks, some disappearing under the weight of a hidden catch. The placid surface belied the events in the deep. The miles of soaking line now held struggling and dead animals, no longer free to roam the ocean. They dangled from hooks set in sixty to ninety feet of water, ready for picking like grapes on a vine.

At 6:00 a.m. we breakfasted on bagels, cream cheese, and orange juice, the meal lasting no more than ten minutes. The race to retrieve the line from the water was on. The longer the gear soaks in the water, the more opportunity sharks and killer whales would have to claim our catch. As Captain Coyle explained, "The longer we take for our breakfast, the longer the sharks have for theirs."

Gasping for air, I finished a raisin bagel just as the *Defiance* lurched forward to reach the first hiflyer. The calm seas and minimal winds made for an ideal retrieval for our untested crew. The hiflyer was rapidly pulled through the port fish door, and the ground line connected to a hydraulic winch. The captain donned his gloves as each of us took our assigned positions. Jack was the runner, carrying and securing all hiflyers and buoys to the bow. Cooters' job was to gaff the fish and spool the empty gangions. I would assist Cooters.

While we waited for the first hook to appear at the surface, Cooters instructed me in the use of the gaff. The old wooden gaffing pole was eight feet long and held a huge six-inch rusted steel hook at one end. He instructed me to reach into the water, over the top of the fish, and plant the gaff hook into one eye; the fish is then hoisted aboard ship. He explained that a properly placed eye gaff lands in the head, preventing damage to the valuable body flesh. He then cautioned that a misplaced gaff can allow a lip-hooked fish its freedom, and a body-gaffed fish can cause huge tears in the meat, dramatically lowering the price at dockside. This is not insignificant if one considers that half the price of a huge fish could spell the loss of five hundred dollars per animal. I became very nervous at the prospect of gaffing, especially after Cooters explained the risks, and debated passing on the assignment, though if I did, the opportunity to gaff such a remarkable oceanic predator would be lost. I strengthened my resolve to try.

The first line surfaced with barely a splash. Successive lines contained either empty hooks or remnants of bait half eaten by unseen animals. Hook number fifteen produced our first day's action. The captain shouted, "Fish on," and the crew dove into action. We stared down into fifty feet of shimmering sea. The torpedo-like shape first appeared small, but as it spiraled upward, pulled by the hydraulic hoist, it grew much larger.

"Small dead sword," the captain called out. He claimed that his special Kmart sunglasses enabled him to see into the deep blue, and he always seemed to read fish before the rest of the crew. Even from fifty feet down, his uncanny ability to identify the catch was remarkable. He tersely announced every fish, which helped the crew to prepare. Words like "green mako" ignited us into action while "small dead sword" allowed us to relax.

The hooked fish broke the surface without moving, and Cooters calmly planted the gaff into the eye. The shape was elegant and graceful, even in death. He walked the sixty-pound creature to the open fish door and pulled it to the deck with extreme care. It was placed on thick deck carpeting to protect the fragile skin. As the crew continued to pull line, I kept glancing at the fish, mesmerized by the beauty of my first swordfish.

Cooters pulled several more small dead swords from the water. The catch was light and disappointing after our first six miles of line, but the good news was that the initial catch was unmolested by sharks. Minutes later, several swords came to the surface in pieces, evidence that our luck had run out. I found the head of an estimated seventy-pound swordfish that I tossed on deck, to be used later for my college research project. The remaining heads were tossed overboard, a visible reminder to the captain of our mounting losses.

The opportunity to gaff my first broadbill came around 9 a.m. when the captain pulled a taut line and called "green medium sword," the term *green* indicating that the fish is still alive and fighting. Cooters nodded to me, laughed, and handed me the gaff as he barked, "Go get 'em, Tiger." My hands shook as I searched the water for the target. The bullet-shaped, graceful swimmer, arching toward the surface, loomed much larger than I'd expected. As with all fish, it panicked when it saw the boat and pointed its head downward towards the safety of deep water. With powerful tail beats, it fought the hydraulic winch in a desperate struggle for freedom. But its efforts were futile as it was forced toward the surface, losing strength with every effort. The lactate buildup in its thick tail muscles caused cramping as its broad black bill smashed the surface of the water. Its huge turquoise-blue eye turned to watch the crew. The remaining seven feet of its powerful body hung vertically just beneath the surface, and its slowly beating caudal fin signaled near-exhaustion. I felt engulfed by its struggle.

"Gaff the damn eye," Cooters yelled.

I planted my feet and placed the hook over its head. Just then, its sickle tail broke the surface, violently splashing water in a last desperate attempt to survive. The giant fish glided forward against a taut line, but I anticipated its movement. I angled the steel gaff at its eye, closed my eyes, and blindly jabbed at its head. This was poor technique from start to finish. Cooters' instruction called for a smooth motion over the fish's head into the eye, and I had watched him perform this movement several times, but now I panicked. The sharp, jabbing hook near the eye had inflicted pain, sending the fish into a frenzy. Her head twisted furiously, striking the gaff pole and causing a tremendous crack and splintering the wooden handle, breaking it in two. I was soaked and white-knuckled, still holding the pole's end.

I stood paralyzed. We watched in horror as this three-hundred-pound gladiator broke free and angled for the deep. Exhausted from the struggle, with a crimson stream of blood flowing from her jaw, the animal glided at a forty-five-degree angle toward the deep, searching for rejuvenation and safety in the cold, oxygen-rich water. Cooters reacted immediately. He disappeared, and then reappeared with a harpoon. It was too late. The fish's shape was still visible at thirty feet, with the jagged gaff still jutting from her jaw. Moments later, she was gone. Captain Coyle was furious. Shouting various expletives, he ranted on about "college boys" and informed me of the exact dollar amount I now owed the boat.

Calculating a three-hundred-pound fish at three dollars per pound, it seemed I owed roughly nine hundred dollars due to my incompetence. I was humiliated and vowed to perfect my skills before being thrown overboard. Cooters resumed the gaffing responsibilities, and I quietly awaited the next opportunity. Perhaps a dead fish would give me an easier opportunity for practice. Cooters read my mind and encouraged me to try the next dead fish.

The remainder of the line produced no further swordfish, although several pests, known as blue sharks[13], were scattered among the lines.

Nearing the end of our set, the line popped tight, and a shark broke the surface. It appeared to be dark blue above, with a snowy white belly, a type of camouflage known as countershading. Color is the first hint of the species. Blue indicates either a valuable but dangerous mako shark, or an inedible blue shark. We scanned the shape for further clues to its identity while Cooters stood poised with the gaff. I grew excited as I made my identification and brazenly called "small blue" before the captain. He paused, frowned, and nodded in agreement, gruffly announcing, "Damn blue dog!" The crew relaxed and Cooters pulled his pliers to cut the line.

----

[13] Blue sharks (*Prionace glauca*) are among the most prolific sharks of the western Atlantic Ocean. Colored a dark cobalt blue on top with a pure white underside, they are graceful, nomadic swimmers that feed on smaller fish and squid. Reaching a size of twenty feet and four hundred pounds, they present a major menace to longlines because of their high fecundity. In fact, the incidental or bycatch of these sharks usually exceeds the catch of swordfish. However, they lack any food value because this shark contains high urea content, a breakdown product of protein that makes them unpalatable. Only their fins are edible and are used in the Asian market for shark fin soup. The fins are procured by a method called "fining," which involves cutting off the animal's fins and tossing the rest of the living shark back into the sea to die an agonizing death. Such an inhumane slaughter of these animals was banned in the United States in 1994, but there is no international ban on this crime, and the practice of fining still represents a serious threat to this stock. Fining kills some 73-100 million sharks worldwide every year. The movie "Shark Water Extinction" illustrates this barbaric practice.

The next clue to the species was the body. A lean thin body indicates a blue shark, which may weigh 350 pounds as a 10-foot adult. By comparison, a 10-foot mako shark would weigh beyond 1,000 pounds. I was pleased with my identification skills and watched the fish with fascination as it was winched to the surface.

Two rusted hooks hung from each corner of the jaw, giving the comical appearance of a handlebar moustache. This shark had been caught before. The line was cut, and our fish watched us cautiously as it glided deep beneath our keel. At a cost of two dollars per Japanese Jinkai hook, we were losing money on each blue shark that we caught, and the cost mounted as we pulled eight more blue dogs.

Approaching the last two miles of gear, a series of empty hooks finally yielded a cry of "small green sword." All hands prepared for capture. The fish approached the surface listlessly. Just as the sword broke the surface, Cooters made a direct hit to its right eye, producing a streak of bright crimson arterial blood over the water. Its shaking head shattered the quiet surface as it struggled to escape the steel barb lodged in its orbit. The pain must have been unbearable, as judged by the degree of violence in its reaction. The little seventy-five-pound fish was pulled aboard, streaming blood, with plenty of life remaining in its tired body. Two heavily-gloved crewmen grabbed the fish by the bill, gently placed him on the carpeted deck, and proceeded to cut the sword from the fish with a hacksaw. The sound of metal on bone reverberated across the deck as the sharp, bony sword was removed to prevent injury to the crew. In less than thirty seconds, its weapon gone, the fish lay quietly in a pool of blood surrounding its head. The scene resembled a murder more than fishing. The animal gasped occasionally, but death came quickly. Almost immediately, the crew was back to the retrieval, unconcerned that we had just taken another life.

The last portion of the set produced a snarled tangle of lines, hooks, shredded bait, and crushed bullet buoys. The twisted mess looked like a hoard of snakes, with cut lines and hooks lost to the sea. Our captain suggested that the probable culprit was a large shark, probably a large mako that had become entangled in the gear. Other possibilities, such as the great white, a whale, or even a manta ray, were suggested but we would never know the answer.

We stored the last of the mangled set and began cleaning and storing the catch. It was nearly 4:00 p.m. when we started to process the catch. Each animal was cut lengthwise at the belly, and guts were removed and examined. The last meal of the majority of swords had been squid and small fish like mackerel and herring. Previous to this, the largest fish I had cleaned were fifteen-pound slammer bluefish, so the experience of nearly climbing into the body cavity to dress a swordfish was remarkable, and these were considered small swords.

After removing the inner organs, we turned the fish on its back, and Cooters taught me the proper way to complete the task. The bloodline is a black streak of blood that runs along the vertebral column inside the fish. It must be cut and stripped before packing the fish. Using ultra-sharp knives, we sliced the thin membrane and removed its jelly-like substance from the length of the fish, then scrubbed along the vertebrae with salt to clean any remaining parts. The next step involved proper decapitation. The heads were tossed overboard and the swords kept for trophies. After each fish was hosed down and washed clean, we began the loading process. Jack and I stood above the hold and carefully passed each fish down to Cooters. He gently placed the delicate cargo on their backs, lining each body cavity with ice. We handled the fish like gold, for fear of harsh reprimand for dropping one.

Cooters talked as we carefully packed the fish. He remembered a story aboard a different boat that occurred while the crew was hastily loading the day's catch. A large specimen of some seven hundred pounds had been hoisted by crane but, due to a poorly placed tail lasso, the animal fell headfirst—fifteen feet—into the hold. The resounding impact echoed through the boat, causing a shudder from stem to stern. The captain, thinking they had collided with an object at sea, panicked and raced around the boat, preparing to call a radio SOS. Reassured by the crew that it was their blunder, not a collision, the ill-tempered captain at first smiled with relief. But after assessing the damage, he went ballistic as he looked at the torn body of this prize swordfish. This so angered him that the crew was sent to their racks that night with a severe reprimand and no dessert, and the devalued fish was deducted from their pay, a perfect example of the authoritarian role a captain plays aboard ship.

We completed the loading in an hour. The deck and gear were then hosed down with fresh water.[14]

———◁◇◇▷———

The catch is kept in the hold until the boat returns to dock three to four weeks later, at which time they are unloaded. But fish can be kept aboard until market prices are optimal for sale. Much like stocks on the stock market, fish prices can vary greatly from day to day, and twenty-four hours may make several thousand dollars difference to a single boat. (Value depends on supply and demand, with the fish buyer helping to advise when to unload the catch.) At dockside, each fish is carefully unloaded by winch, weighed, and inspected for quality. Swordfish are first checked for temperature; a thermometer is inserted into the flesh to confirm the fish has been well chilled. The ideal temperature is 32 degrees Fahrenheit, but values between 32 and 35 degrees are acceptable. A temperature greater than 40 degrees is considered suspicious for spoilage. The tail is cut, and the fish then is inspected for color, firmness, smell, bloodline, and skin marks. At dockside, this cut of fish is called "Boston cut," meaning no head and no tail.

The fish is graded as one of three qualities: top/diamond, maroon/burgundy, or brown bloodline. Diamond-graded fish have no skin bruises or

---

[14] Saltwater rusts everything it touches, and we were constantly reminded to hose everything before storage. This fact struck home as I packed the last bullet buoy and noticed an old rod in the forward hold. Rust encrusted the reel, and it seemed evident that it had never been hosed or properly oiled. I retained this image in my mind and, from then on, made a point always to properly rinse all our gear.

marks, silvery colored skin, and a bright red bloodline in the meat adjacent to the backbone. The next level of grading is maroon or burgundy, referring to a dark red bloodline and indicating a longer period of storage than the diamond. The lowest-grade fish is called a brown bloodline. Although possibly still fresh, this grade indicates an unknown time of storage (probably because the blood has turned brownish in color by this time). Price differential between grades may be huge, with a diamond possibly valued at five dollars to six dollars per pound and a brown bloodline as low as one dollar to two dollars per pound, translating into a three-hundred-dollar difference for one swordfish weighing more than one hundred pounds. When the fish are graded, the tail section is severed and discarded, and the remaining portion of fish is placed in a box on ice, always on the same side, labeled the "bad side." The good side is shown to the buyer and is well protected throughout shipping.

But price also differs according to size. Fish weighing 100 to 250 pounds or more in dressed weight are termed "markers" and fetch the highest prices. Restaurants seek out these markers because they can serve uniform-size steaks to their customers and have fewer odd-size pieces to deal with. The fifty-to ninety-nine-pound dressed-weight carcasses are termed "pups." They offer less uniformity. The twenty-five-to-forty-nine-pound dressed-weight fish are called "rats" and generally are not chosen by chefs.

Once the swords arrive at the local fish market, they are stored on wet ice and then cut to order. They may remain on ice for up to five days before being sent to the freezer. An average fish market may buy five hundred pounds of fresh broadbill per week, and with such a high market value, none goes to waste. The fish kebobs or end cuts are sold at a reduced price. Once brought to plate, the swordfish steak looks like fresh pork. Its subtle-sweet flavor and texture appeal to a wide variety of palates across the globe.

But there are big problems—and no succulent fish to enjoy on the plate if the fish is not properly iced and refrigerated. Fresh fish begin to spoil immediately after the animal dies due to a biochemical process involving bacteria, enzymes, and chemicals. The majority of spoilage, however, is caused by bacteria that normally are well controlled by the host's defense system but which now invade the flesh and begin to multiply, breaking down the flesh and creating a fishy odor caused by amines. Enzymes continue to work by breaking bonds and softening flesh. The chemical action of oxygen on unsaturated oils also causes rancidity, particularly in fatty fish such as salmon and mackerel. However, all of these changes can be delayed by maintaining fish at very cold temperatures.

On rare occasions, three-to-four day old, well-refrigerated swordfish can become 'jellied,' with no noticeable change in odor or appearance. The texture of the flesh softens, probably due to bacterial activity, and the condition is not obvious until the fish is prepared for cooking. Though unpalatable, eating 'jellied' swordfish has not been reported to be harmful. However, these catches are not marketable.

One risk of eating improperly refrigerated fish is the threat of scombroid poisoning, an intoxication caused by the protein, histamine. Fish most commonly affected are of the scombridae family, including tunas and mackerels, plus bluefish and dolphinfish. The U.S. Food and Drug

Administration (FDA) reports that 50mg/100 grams of histamine in tuna have the potential to cause illness, although higher levels of histamine are noted in actual illness. Signs and symptoms may be reported as early as two minutes after ingestion and include sweating, nausea, headache, and a burning or peppery taste. But this type of food poisoning can be avoided altogether if fish are properly stored in ice and kept at or near freezing until they hit the grill.

To avoid the threat of food poisoning, there are several ways for consumers to determine the freshness of the catch, and the single best way may be the sense of smell. Good fish smells of fresh seaweed, not unpleasant at all. The first area of foul odor would come from the backbone, between the gills and the body cavity. However, the ideal fresh fish has been de-gilled, as this is where spoilage begins. The eyes, if present, should have a shine, a convex shape, and clear corneas. Sunken eyes signal a fish that is no longer fresh.

Although fish generally are safe to eat, there are other hidden dangers that should be recognized. Toxic metals and parasites always are possible contaminants, although unusual in today's modern seafood industry. The highly migratory species are so-called because they feed at the top level of the food chain, unavoidably accumulating numerous poisons from the environment, including plastics, pesticides such as DDT, and metals containing mercury. Swordfish was banned in the 1970s when unacceptable levels of mercury were discovered, causing a meteoric drop in consumption. Some mercury is released naturally into the environment, but an equal amount can be released as a result of human activity such as burning coal and other fossil fuels. At excessive levels, mercurial poisoning can cause numbness, a stumbling gait, and, at the extreme, coma and death. The FDA reestablished acceptable standards for methyl mercury to be one part per million, ten times lower than the lowest levels known to create adverse effects. Experts now recommend eating apex predators, particularly swordfish and tuna, no more than once per month for men; women and children should avoid them altogether, considering the risk of mercury poisoning. Although the average level of methyl mercury in swordfish is very close to the edge, a whopping 0.88 ppm, it should also be noted that there is no evidence of actual illness from these stocks, only speculation.

————⟨⟨⟨⟩⟩⟩————

Back on deck, we watched the billowing white canvas of a small sailboat on the horizon. It was an eastbound thirty-eight-foot Morgan, and I jumped to the wheelhouse to see if we could make radio contact. It seemed that the captain had the same idea, as they already were communicating via the radio. The boat's husband and wife team were outbound for Bermuda, anticipating a five-day trip to cover the six hundred miles to the island's western shore. They discussed the favorable weather report and a sighting of whales, bid us good fishing, and passed us by 4:00 p.m. The smell of roasted chicken wafted up on deck as I headed for dinner, another day of fishing under our belts.

CHAPTER 7

First Set Gets the Tuna

Following dinner and clean up, we prepared to set the gear, while the captain disappeared to his sanctuary, planning our next location. The boat's engines started, and within the hour we were back along the Hudson Canyon's west wall. But this time we had company, to the great disappointment of our captain. The first boat arriving at the scene has what's called the prime set, meaning that they get priority as to where to place their gear. It was our obligation to give them room in order to avoid entanglements, and it's this protocol that forced our boat into a less productive stretch of water to the south. They chose the west wall along the one-hundred-fathom curve where presumably they would get the majority of the fish.

Knowing that swordfish typically swim inshore for evening feeding gives the first boat on scene the best chance at success; we had missed our opportunity, and the captain was disappointed. But all was not lost. The prime set boat, out of Point Pleasant, was nearing her last day at sea. Both captains agreed that the *Defiance* could move into her prime set location in exchange for milk and groceries for the depleted inbound boat. That decided, we continued to steam southeast for another hour.

Setting the gear proceeded in the usual fashion but the mind-numbing repetition was especially boring due to a calm Atlantic. Searching for entertainment, I picked up my dog-eared copy of *Moby Dick*, but couldn't concentrate. We were now drifting one hundred miles from home and, after six days at sea, had not earned a dime. Although I had not embarked on this trip for money, I began calculating pounds and debt. I realized how many meaty little torpedoes we needed to kill before I had money in my pocket. We now had five thousand pounds in the hold but we needed over fifteen thousand pounds of fish to show a profit. Overcome with excitement, I scribbled numbers with pencil to paper, hungrily calculating my potential gain.

At roughly ninety pounds per average fish, and approximately three dollars per pound, we still needed to boat close to one hundred more swordfish before we were in the financial black. This seemed an easy order, considering that we had two more weeks of fishing. However, I failed to anticipate three conceivable problems. The first and most common dilemma would be the difficulty in finding fish. The second was the weather, which can be moody and uncooperative in the western North Atlantic. And the last was the constant potential for problems aboard ship. At this point in time, we had not found many large fish, but the weather had held and the *Defiance* was running smoothly.

The next morning, after an uneventful watch and a solid two hours of sleep, I felt renewed enthusiasm. I had started my journal (unaware that it would someday become the impetus for a book), and I had a fresh, intact swordfish head for my college research project. Coming out on deck, I scanned the horizon for ships. We were alone on a calm, sunny morning, with only a few circling birds to keep us company. I watched one alight on the deck, resting from its long sea journey.

The captain identified these quick little black birds as storm petrels, but I pulled out my Audubon book of seabirds, just to be sure. The book identified the bird as a sooty shearwater. He remained still, its tiny black eye following my limited movement. A bright idea dawned on me: why not bring it closer by feeding it. I grabbed some baitfish and tossed it, but it landed too short, just as the bird took flight. Several more birds flew above and I wasted more bait to the sea. They scooped up the scraps and continued to circle the boat as we prepared to retrieve our gear.

We felt cheerful and full of optimism that this would be our day. When the last of the gear was hauled by noon, however, most of the hooks were empty and the bait mangled. Our disappointing tally amounted to six half-eaten swords, a large number of live blue sharks, and ten small, unmolested swords weighing a total of about one thousand pounds. The poor catch served to dampen our spirits once again.

Also among the set were several albacore tuna (*Thunnus alalunga*). Each measuring about two feet long and weighing about thirty pounds, they were much smaller than their family members, the famed bluefin and yellowfin.

They can reach four feet in length and one hundred pounds, but all eight albacore together amounted to a lowly 250 pounds.[15]

We cleaned up, then steamed west to keep the rendezvous with our mates who were fishing inshore along the canyon's western wall. As planned, we were to find the other boat and provide groceries as a trade-off for their prime-set fishing location. Arriving at the western wall by 3:00 p.m., we neared the vessel, curious about our competitor. We noticed that the other vessel, about the same length and shape as the *Defiance*, was cleaner, and bore a fresh coat of red paint. Three bearded and shirtless men stood on deck in their orange oilskins, all well-tanned, except for their foreheads, which remained pale under various types of caps, giving them what the offshore fishermen call a "tuna tan." The crews acknowledged one another and the captains finalized plans via the radio for the transfer.

As the two vessels angled closer into position, the strangers eyed us with equal interest; we were probably the first human contact they had had in weeks. By comparison, our crew looked tired, ragged, and dejected, due mostly to our empty hold. None of us had shaved for a week and certainly none had bathed. We were, in the truest sense, a motley crew. The other fishermen seemed bemused by our miserable appearance, grinning at us as we eyed their catch in disbelief. Beautiful bigeye tuna, most of which

---

[15] The albacore generally are less valuable due to their tendency to dry out during cooking. Also known as "Charlie the Tuna," this is the fish found in cans as white meat tuna.

appeared to weigh two hundred pounds or more, lay across their deck.[16] The men had just completed their pull, the last fish still pounding the deck with powerful smacks of its tail. This day clearly belonged to the crew from Point Pleasant, not to the *Defiance*.

———⟨⟩———

Tuna, including yellowfin, bigeye, and bluefin, are handled in a different manner than the broadbill when caught. Because many of these fish are transported to Japan for sushi, they must receive additional care in processing; each fish is bled through a cut to the pectoral fin artery. Another cut is made through the tailfin or caudal artery, and the gill plate is lifted. Then the artery under the gills is cut, and a deck hose is used to pump water and blood through this arterial system, bleeding the fish through the tail. After the head is severed, the final step involves pushing monofilament fishing line through the vertebral column to the tail, destroying nerve

---

[16] The bigeye tuna (*Thunnus obesus*) are large apex predators that range from Nova Scotia to South America. They are valued as sushi due to their high fat content and may reach four hundred pounds in weight. Unique for their big eye that enables them to see well into the depths, they feed farther out than their cousin, the bluefin, and are smaller and easier to catch. Also, because of their culinary value, captains consider the bigeye prime targets, seeking them out along the five-hundred-fathom curve of the Hudson Canyon. They are currently listed under ICCAT as vulnerable, a category defined as likely to become endangered unless the circumstances that are threatening their survival are reduced or improve. This category has been listed under the IUCN (International Union for the Conservation of Nature) and extends along a spectrum of "least concern" to "extinct."

endings and preventing muscle twitches that may increase spoilage of the meat. A general rule with fish is that the "briefer the battle, the better the fish." As tuna struggle, they build lactic acid in their muscle, giving it a burnt taste making it very undesirable for sushi.

The captains skillfully maneuvered the boats into a bow-to-bow position, preparing to transfer supplies. This type of cooperative interaction is common among longliners, as it sometimes becomes necessary to depend on other boats for groceries, gas, or water. Each boat, supplied with limited resources, is also subject to fire, sinking, mechanical breakdowns, and crew meltdowns, and such vulnerability forces fishermen to cooperate at sea. It is like a small but dysfunctional family, with crews competing but also relying on one another.

Once the boats were positioned twenty feet apart, Captain Coyle announced over the PA, "Ready for transfer." Eye to eye with these bearded Vikings, their skin covered in fish blood and scales, I stood like a fool, motionless and unsure of my assignment.

"Toss it, man," Cooters yelled from behind as he handed me a dripping plastic gallon of milk. All that was required was a fifteen-foot toss across rolling seas, onto a pitching boat. Why was I hesitating? I made a distance judgment and then hurled the jug with my best bowling ball pitch, watching as it arched over the water and landed on the slippery deck with a thud. The first item had been safely transferred. The remaining goods included two more gallons of milk, a carton of cigarettes, one iced six-pack of Budweiser, and five pounds of frozen lunchmeats. Our part of the exchange was complete.

"Let's move it," Captain Coyle anxiously ordered over the PA.

I heard concern in his voice. He wanted more space between the vessels. The engines grumbled as we slipped backward several hundred yards, then turned northwest to start our set. Both crews waved, and the captains exchanged final messages just as the other boat pulled away to the west on a course for home. We were encouraged by their success, landing over thirty-thousand pounds of mixed sword, tuna, and mako shark. These waters were bountiful, and it was now our turn to get our fair share. We steamed six miles due northwest to begin our search for another opportunity to fill the hold.

During our supper of barbecued spareribs, baked beans, corn on the cob, and apple pie, the captain appeared unusually cheerful. He regaled us with stories of the sea and, as usual, about his favorite subject, the bluefin tuna (*Thunnus thynnus*).[17] One disturbing story involved that of a longliner fishing the Grand Banks off Newfoundland, which had caught a tuna estimated at

---

[17] Many captains prefer to avoid the bluefin altogether, due to the animal's extreme size and the labor-intensive effort required to boat them. One of the largest and fastest predators of both tropical and temperate oceans, this member of the mackeral family can weigh in excess of fifteen hundred pounds (the largest ever caught off Cape Breton, Nova Scotia) and can cross the Atlantic Ocean in less than fifty days, as measured by tagging studies. They tend to migrate south to north, along the western Atlantic, arriving off the shores of Canada in June and remaining there through October, then returning to winter in the Caribbean.

nearly a ton. After attaching it to the hydraulic winch for boating, and hours of struggle to hoist the giant animal onboard, seas began to build ahead of an approaching storm. The crew decided to quit. The behemoth simply was too heavy to be boated using conventional longline gear. The giant fish, now dead from exhaustion, was dumped back into a frothy black sea. This

---

Bluefins hunt in schools, grouping with others of like size; and the schools are often small, consisting of less than fifty fish feeding on smaller schooling fish of herring or mackerel, squid, and capelin. Fishermen have grouped these schools into commercial categories consisting of young school (less than 14 pounds per individual fish), school (14 to 66 pounds), large school (66 to 135 pounds), small medium (135 to 235 pounds), large medium (235 to 310 pounds), and giant (over 310 pounds). The animals swim in two different types of school formation, depending on activity. They may swim side by side, called a soldier formation, or in an arc, called a hunter formation. For protection, bluefins sometimes follow smaller tunas, hiding beneath the school to avoid detection from enemies. Because smaller tunas are skittish, they serve as guards and warn the bluefins to the presence of predators.

One of several adaptations giving the bluefin their great swimming speed is a physical attribute, unique to most oceanic pelagics, termed a countercurrent exchange mechanism. It is this very system that has ensured their evolutionary success over millions of years. Through a rete mirabile, or "wonderful net," warmed blood coming from contracting muscles transfers heat to overlying veins, providing and conserving body heat and elevating the body temperature some ten degrees higher than the environment. In poikilothermic or cold-blooded animals, warm muscles are essential for power and rapid food digestion, resulting in faster swim speeds to avoid predators and to catch prey. Amazingly, the bluefin can swim at speeds of forty miles an hour and accelerate an extra twenty miles an hour, corresponding to a land speed of nearly three hundred miles an hour. Because bluefin are hydrodynamic, they encounter minimal drag, an evolutionary adaptation that has afforded them a speedy lifestyle.

reproductive, thirty-year-old fish surrendered its life for no reason, a great waste due to greed.

However, it's not just about good storytelling. The bluefin tuna are in the same precarious situation of being overfished as the billfish and sharks. Historically, the bluefin has been fished on a large scale since 1956, when longlining for tuna began in the warm equatorial waters of the Atlantic Ocean. By 1965, several hundred boats were setting out millions of hooks, causing a steady decline in stocks. Further depleting the stocks, the Japanese longlined the Gulf Stream during the winter months of the sixties and seventies, catching up to ten thousand fish per year over the two decades. By 1974, both the National Marine Fisheries Service (NMFS) and the International Commission for the Conservation of Atlantic Tunas (ICCAT) stepped forward and began setting catch quotas, but the battle between conservation and fishermen continues to this day over the highly prized bluefin.

Another factor leading to a successful fishery was the advent of zappers in the 1980s. Zappers are an electrified harpoon created by an aeronautical engineer from Chatham, Massachusetts. The zapper immediately stuns and then kills the fish by sending electricity through its body. Although this may seem a painful way to die, many argue it is more humane than letting a fish struggle for hours before it exhausts on the line, a method that left many of the ultra-giant bluefins to die without ever being recovered.

The reproduction cycle also may play a role in the scarcity of bluefin tuna stocks. The two major spawning grounds in the western Atlantic are the subsurface waters of the Straits of Florida and the Gulf of Mexico;

there, spawning takes place during April, May, and June. A female is not reproductively mature until she reaches about four hundred pounds, when she may release sixty million eggs into the water during spawning season. The male then fertilizes the eggs externally by a reproductive strategy called 'oviparity.' Once an egg is fertilized, the chances of it surviving are extremely minute. Potential predators include whales, crabs, fish, and nearly anything larger than its 1.1-millimeter size. Although sixty million eggs sounds like a large quantity, the chance of a hatchling living to adulthood, over the course of eight years, is a staggering *one in forty million*. This translates to two grown females producing three adults from the efforts of an entire spawning season. But by six years of age, this three-hundred-pound creature has few enemies. If the animal is not caught or eaten, it is thought that it can live to almost thirty years of age.[18]

We returned to work, but it was difficult to shake the graphic image of a giant bluefin, hanging limp and dying, then dumped back into the ocean like yesterday's garbage. We completed the evening set in light seas and gentle winds. The crew was revitalized by the calm weather and the potential for a better catch ahead. I turned in, reading *Tales of Whales*, just nodding off when the familiar knock on my rack brought me back to day seven at sea.

---

[18] In addition to man, natural enemies of the adult bluefin tuna include sharks, orcas, pilot whales, and perhaps even sperm whales. Although records exist of sperm whales catching swordfish, it is highly unlikely that a sixty-foot, relatively sluggish whale could catch a healthy swordfish, although we lack much information on the interaction between swordfish and sperm whales due to their deep-sea existence.

CHAPTER 8
Pothead

A dim sunrise gave way to clouds and heavy rain. A squall line had formed to the east, its winds whipping the sea into six-foot froth. The slippery deck created an unstable platform on which to work, and the sea spray blew steadily into my face as rain pelted my jacket. I went below to don extra gear and grab a last cup of coffee. The radio crackled with local boat traffic, but at first I could see nothing moving, only the set drifting close by on an otherwise empty ocean. To the west, I could make out the shape of an oil tanker, less than two miles distant and barely visible through the dense fog. It appeared to be steaming due west, heading to New York Harbor. Remembering the captain's words, I realized that we were not far from harm's way. In pitching, rolling seas, shrouded in fog, it would be possible to collide with an obscured vessel, especially if the two were less than a mile apart.

The engines started, and we glided up to the first hiflyer. Initially, the hooks breaking the water were empty, but we noticed a heavy tension on the ground line. The pitching deck groaned as each foot of cable begrudgingly wound back onto the drum, the strain increasing with every increment of the turn. We stared intently into the black water, waiting to see the cause of this tremendous weight. Through a torrential downpour, the captain bellowed, "Pothead!" I felt annoyed at his inappropriate timing to find fault with the crew's personal habits and recoiled at his abuse; Cooters witnessed my reaction and laughed. He pointed to the frothy sea surface and the huge, struggling-black shape appearing just beneath the waves. The captain again called "Pothead," and it dawned on me that he was referring to the creature in the water and not to the crew. I searched my memory for such a term. "Pilot whale, you dumb ass," Cooters barked.[19] I breathed a sigh of relief and studied the water.

Two hundred yards off to port, we watched the source of the tangled lines. The whale was nearly twenty feet long. The poor trapped creature was fighting for its life. Each time its rounded black head would break the surface, it was pulled below again under the weight of our gear, growing more tired with each attempt at breathing. The huge black flukes of the pilot whale were almost touching the boat as its tail slashed at the water, its immense size and strength frighteningly close to the gunwales. Captain Coyle winched the whale as close to the *Defiance* as possible, tugging on the ground cable that ensnared its tail. The whale smashed against the line

---

[19] The pilot whale (*Globicephala macrorhynchus*), also called pothead or blackfish, is a jet-black, toothed whale that frequents the western Atlantic. They may reach a weight of three tons, grow to twenty-plus feet in length, and can live up to fifty years of age. Identified by a bulbous forehead, from which the name pothead is derived, they feed on squid and fish and are often seen at the surface in small groups of four to six individuals. The killer whale is their primary enemy, and potheads have even been known to beach themselves to avoid being devoured.

and ripped it from the captain's gloved hands. With rolling seas making stability difficult, the twisting, rolling, panicked, two-ton animal made the situation extremely dangerous. Keeping its head submerged and heavy flukes in the air, the whale continued to smash the water, anguished and squealing with frustration. Finally, after what seemed like hours of struggle, our swearing captain was able to cut the cable. He had watched tail-wrapped whales become entangled with the propellers during similar interactions, sometimes with disastrous results, and was determined to free this animal. "Not on my boat," he yelled as he untwisted the last of the line from the great black flukes. It was a glorious moment when the whale left us with a loud slap of its tail against the hull, then dove beyond sight.

We held our breath and waited, eager for reassurance that the animal was well. Silently, we scanned the seas for telltale signs of life. Finally, after a long thirty seconds, the huge black form broke the surface only a few hundred yards from the stern and leaped from the water, celebrating its freedom. Although perhaps an anthropomorphic viewpoint, the whale seemed thankful to be released, and we sensed its gratitude for our efforts. Four times the whale repeated this feat for us within full view, and we were elated by the performance. Within moments it disappeared under the waves to rejoin the pod. Although brief, this encounter had created a bond between the whale and crew; I think we all wished it well.

———◆———◁⬖⬗⬖⬗▷———◆———

Encounters between pilot whales and man are common in the western Atlantic. These cute little whales are interesting and fun to watch, but since a two-ton whale needs about one hundred pounds of fish per day, they often feast on a longline catch. Stories of fishermen fighting back with guns, to deter the whales, have been documented. Although this practice is now clearly illegal, thanks to the Marine Mammal Protection Act of 1972, the frustration of losing tons of fish to a whale can be overwhelming. At times, miles of lines have been pulled to the surface holding nothing but swordfish heads by the dozens, amounting to thousands of dollars of financial loss.

Fishermen claim their gear is not endangered when a pilot whale pod is physically spotted. It is the invisible animals that are reputed to be the real threat. These animals presumably wait, hidden deep beneath the waves, and follow boats for this purpose; the opportunity for an easy meal is just too appealing. Although human stealing is immoral, the law of the sea is to survive, and the logic of an easy lunch with little caloric expenditure makes sense.

Stories of pilot whales associated with the presence of large sharks have also been well documented. I was fortunate to observe a pod of free-swimming whales only thirty miles offshore in the summer of 1995. We were sportfishing in a shark-tagging tournament in an area known as Glory Hole. The seas were heavy, but the boat, a fast, forty-five-foot ocean yacht, reached the grounds early. The chum bucket was placed overboard and the baited lines tossed. After several hours and no action, I climbed into the tuna tower to observe our surroundings. Suddenly, over the sound of slapping waves and the hum of the generator, we heard a tremendously loud hiss, like an exhale, and noticed a pilot whale swimming directly off the stern, perhaps five yards from the transom.

It swam on his left side so that it could watch us, its black eye the diameter of a coffee mug. Then, apparently disinterested in our boat, it glided past and dove for deeper water. The fish finder recorded the remainder of the pod, only sixty feet beneath our keel and some eight strong. We were stunned and elated at this wonderful encounter. We didn't, at that time, realize that pilots often swim with other pelagics. Just minutes later, we hooked a ten-foot, one-thousand-pound mako shark that had been following the pilot whales.

The relationship between pilot whale and shark is not clear, although several theories have been presented. Both the pilot whale and the mako shark feed on similar prey, such as mackerel, squid, and herring. Another possibility is that the mako was following the pod of whales in search of wounded or young animals. Whatever the reason that sharks follow the whales, the threat created by a half-ton mako following a pod of whales paints a dynamic picture of the predator/prey relationship in the deep oceans.[20]

---

[20] Another whale that shares the pilot whale's predilection for decimating longlines is the killer whale. Reaching thirty feet in length and traveling in huge pods, these animals actually follow boats, stalking the lines. Although we were fortunate to avoid interaction with these fabled pirates in tuxedos, fishermen in northern waters off the Grand Banks have frequent encounters with the marauders. Stories from the Bering Sea tell of solitary scouts, which are seen, then vanish, returning later with the entire pod. These intelligent animals have apparently made the association of boat with food, creating a perpetual problem for fishermen.

The catch for the day was immensely disappointing; the tally came to seven small swords, six swordfish heads, and several blue sharks. Several smaller sharks, known as dusky sharks, were boated dead. These small, six-foot fish have no food value and were unceremoniously dumped back into the sea, another common waste of marine life known as bycatch. The darkening skies and poor catch matched our low spirits, and I decided that comfort food would help. I headed to the galley to fix burgers for the crew but chose tilefish for myself.

I had chosen a fish from the cooler and cooked it seemingly to perfection. The crew devoured their burgers, but for some reason were looking at my fish as if it had returned to life. Distracted by one of Jack's stories, I turned back to my plate just in time to see an alien life-form coming out of my fish. The little beast was white, thin, and could best be described as a worm. He appeared to be agitated, probably due to the 300-degree cooking temperature, as he twisted out of his hiding place, my dinner. We watched in fascination as he danced a jig on my plate while the crew roared with laughter.

"Do you want some fish with your parasite?" Cooters prodded, almost choking on his burger.

But the poor little critter got his just desserts in the decompression chamber, also known as the microwave oven. I cooked him and the rest of my fish for two full minutes, then smeared butter over the top before sitting down to eat both the fish and the parasite, savoring every sweet bite.

Although minimal and not well known, the threat of parasites is another risk to fish consumers. Most of these worms, however, cannot adapt to human hosts, are generally not transmissible to humans, and are easily killed by proper cooking or freezing. In fact, fewer than thirty cases are reported per year in the United States. To protect consumers from purchasing raw fish that might contain parasites, the FDA provides guidance under its Parasite Destruction Guarantee. This mandate states that fish intended to be consumed raw must be "frozen and stored at a temperature of -20° C (-4° F) or below for a minimum of 168 hours (seven days)." The FDA, in concert with the National Marine Fisheries Service (NMFS), also sponsors an inspection program. In addition, fish wholesalers pay inspectors to perform dockside evaluations of fresh seafood, giving consumers another reason to buy American-caught seafood, as the catch methods and chain of custody of imported products is unknown.

Dessert was strawberry shortcake, made with fresh fruit. Throughout the entire trip, we always had fruit onboard with the exception of bananas. For unexplained reasons, seamen never kept them onboard. According to the captain, whose stories covered a wide range of topics from the mundane to the ridiculous, sailors are a very superstitious lot. For example, a crew should never leave port on a Friday (the day Christ was crucified), whistle on board, leave a hatch cover upside down, shave on board, or put the left foot down when stepping aboard ship. Also, a naked woman aboard a boat calms the ocean (unfortunately we had no naked women aboard, so this may have accounted for our weather difficulties), porpoises swimming around the boat bring good fortune, and a silver coin placed under the masthead ensures a successful and profitable voyage. We chose to question none of these.

One of the most interesting beliefs held by seamen is that it is useless to fight the sea if one falls overboard, which may account for the fact that most sailors can't swim. Two centuries ago, almost a third of all British seamen died while at sea, either from life aboard ship, because of storms, or drowning. Ships sailed at the mercy of the winds, so in the event that a seaman accidentally fell overboard, it may or may not be possible to slow the ship or turn the vessel in time to save him, if he could be found at all. Seamen depended upon experience and luck to stay alive. Should they go overboard during a storm or shipwreck, there were no lifeboat services to rescue them, and, in fact, until the mid-nineteenth century, mariners received less assistance than salvagers. British law protected cargo rather than crew. Drowning was considered a natural event, and once a man was overboard, he already was claimed by the sea. Sailors actually watched as their own shipmates went down, doing little-to-nothing to rescue them. So why learn to swim?

CHAPTER 9
Contemplating Murder

After dinner, I tapped ol' Betsy and took a cup of hot coffee on deck. I thought about our success with the whale and of other giants of the sea like great whites and mighty sperm whales and realized how fortunate I was that I could be part of such a rich experience. But the truth of the matter was that my worst monster, and greatest enemy, was aboard ship.

We had completed the evening set and were drifting at the edge of the west wall. The storm had passed, and now a full moon illuminated a calm sea. The stars were brilliant. It was Saturday night. I imagined my friends having a great weekend while I was at sea. Although homesick, on this particular night there was no other place I would rather have been. Sure, I was isolated, but certainly not alone. Here, on the west edge of the Hudson, weekend sport boats were everywhere, like ants around a picnic. I felt a twinge of envy, as these weekend warriors would resume their normal lives by Monday morning. Meanwhile, I had barely completed my first week aboard ship.

The masses were targeting tuna and sword, and our outspoken captain viewed these little sport boats as the scourge of the ocean. At a ratio greater than fifty sport boats to every longliner, these small vessels occasionally would tangle themselves in our gear. Tonight, there were boats in every direction. The radio buzzed with their conversations, and as I climbed into the dark wheelhouse I could hear the captain cursing these "mosquitoes" under his breath. Most boats had caught little of interest; several had small tuna, but one boat, some four miles east, was reportedly embroiled in its second hour of battle with a monster fish.

The crackling radio reported the fish as "down deep," then nothing but silence, long moments of breathtaking silence, as they waited for the fish to appear. We were like kids listening to a baseball game, waiting for something to happen. I secretly hoped the fish would escape. We listened for another twenty minutes when finally, in a highly excited voice, the captain announced a four-hundred-pound sword near surrender at boat side. Again there was silence, followed by loud expletives, then the words, "Fish off." I had gotten my wish. Anxiety gone, I turned in, falling into a peaceful sleep.

I napped for a full hour but woke around 10:00 p.m., still achy and bone tired. Time for a snack while Jack was still on watch and I had two hours to kill before my turn. The galley was deathly quiet except for the steady hum of the generator and an occasional lapping wave. Mutt slept in the corner, and I hoped he wouldn't wake. I wasn't in the mood for another nip and retreat battle. I was exhausted, sleep-deprived, and had no intention of tolerating his assaults any longer. This small dog had become my nemesis. It was then that I decided to kill him.

The plan was hatched in the same way that I now built my ham and cheese sandwich—slowly and methodically. We were the only sentient beings awake on the boat. It was just Mutt and me. My plan was to step outside and allow our mascot to follow. I then would quietly open the fish door and wait for the inevitable attack. As he stalked me to the open door, I would kick him overboard. The crew was fast asleep and would never know what happened. It was the perfect plot. He would be shark bait. The crew simply would think he fell overboard. The incident would be viewed as an accident at sea and then forgotten.

The plot seemed perfect, but I felt ambivalent about its execution. On the one hand, here was a vicious mongrel that tore every item of clothing I owned and incessantly intimidated me at every turn. On the other, I was reluctant to be so cruel and end a life in such a fashion. But, being pushed to extremes, I wondered if I could tolerate another two weeks at sea with this hellish beast.

A second deterrent to my plan, however, was the potential risk to myself. If I miscalculated, I could be the victim to fall overboard, not Mutt. My oilskins quickly would fill with water as I drifted away from the boat; my screams of distress would go unheard, and I would wander into the black void, alone. This was not a death I would choose for myself, and even this canine ocean-going monster deserved a better fate.

I moved through the door, to the deck, and he followed me outside, right on schedule. But I couldn't do it. I couldn't open the fish door. I knew in my heart that I could not commit such a heinous act. My guilt-ridden conscience would not allow me such a luxury. Seeing that nothing was happening, the drowsy dog ambled back into the galley, too tired to even nip. Just as I turned to follow him, I was rewarded by the sound of dolphins.

———⊂⟨⟩⟩⊃———

Atlantic bottlenose dolphins[21] (*Tursiops truncatus*) habitually follow boats at sea, and the *Defiance* was no exception. The Atlantic bottlenose

---

[21] Atlantic bottlenose dolphins belong to the order of Cetacea that includes all whales; the suborders are Odontoceti (toothed whales, including the killer whale and the mighty sperm whale); Mysticeti (baleen whales, including the blue, humpback, and fin); and Archaeoceti (a presumed extinct group of toothed whales). Well-known for their intelligence, the dolphins' sense of hearing is highly developed, compensating for their poor visibility in their environment. Most sound reception takes place through the lower jaw, which contains a fat-filled cavity that efficiently conducts sound to the auditory center. A small external ear opening behind each eye also is believed to play a minor role in detecting sounds of lower frequencies.

Dolphins also produce sounds for communication as well as for navigation, and current theories suggest a tissue complex in the nasal region as the site of all sound production. The variety of dolphin sounds can resemble whistles, trills, grunts, and creaking doors. (The toothed whales also possess the ability to echolocate, using sound to "see" objects underwater.) Sound is emitted, and the returning echoes are interpreted to determine the size, shape, and distance of objects in the water.

A high-speed carnivore, the dolphin can swim twenty miles an hour, and they've been known to routinely dive to 150 feet for squid and fish. In fact, although not routine, one trained dolphin was able to dive to over seventeen hundred feet, displaying the mammalian diving reflex unique to all marine mammals. Such a reflex allows the animal to conserve oxygen and remain submerged up to ten minutes while vigorously hunting. They may eat up to 5 percent of their body weight per day (a total of seventy dolphin is the most abundant dolphin species along the coastline of the Eastern United States. From Cape Cod to the Gulf of Mexico and as far north as the cold waters of Nova Scotia, south to the warmer waters of South Africa, they inhabit the open oceans (pelagic) as well as inshore harbors and estuaries.[22]

During the summers of 1987 and 1988, nearly seven hundred dead or dying dolphins washed up on various eastern beaches due to what was thought to be a naturally occurring red tide event. But this theory was replaced in 1994 with a more ominous interpretation. Tissue cultures taken from a significant sampling of dead animals showed evidence of morbilliviruses, including the measles virus, which led scientists to speculate that the animals might have succumbed to this human virus. The dolphins have no natural immunity to human viruses and therefore would have no natural defenses. However, it is unknown which human viruses directly are transmissible to marine animals; therefore, this remains speculation. The potential of human to animal viral transmission is frightening and an issue which must be investigated because our human population continues to increase, continues to populate our seacoasts, and continues to pollute our pristine oceans, possibly putting all marine life at risk.

---

pounds) to maintain their high metabolic rate, and can live to nearly fifty years, barring illness or predation by large sharks or killer whales.

[22] There are two ecotypes (forms) in the northwestern Atlantic Ocean: The inshore, known as the coastal ecotype, includes smaller dolphins that frequent warmer, inshore waters. The offshore ecotype is a larger animal, better adapted to colder, deeper waters.

Today's number of bottlenose dolphins is unknown, although estimates now place the population at a healthy twelve thousand animals in the western North Atlantic. Although man does not directly hunt dolphin in the western Atlantic, in many parts of the world they are killed for meat and oil. Dolphins also have been adversely affected by habitat destruction, boat traffic, and pollution. In the Pacific Ocean, however, the current main threat to both the spotted and spinner dolphin is their ensnarement in tuna fishing nets. Millions have been killed over the last few decades in this accidental manner.

In 1949, the Inter-American Tropical Tuna Commission (IATTC) was formed, their goal being to conduct research on tuna populations and to recommend conservation measures. After conducting focused research covering eight million square miles in the eastern tropical Pacific, a major tuna fishery, IATTC's early statistics were astounding. They reported that by the early 1970s, three hundred thousand or more dolphins per year were dying from being crushed or drowned in the nets. Because dolphins are mammals, they need air to breath and can die quickly when caught in nets. This staggering loss of life due to bycatch was disturbing to fishermen, and early interventions aimed at reducing the numbers quickly were implemented. One solution was to allow dolphins to escape the closing nets; experienced European purse seine fishermen taught other fishermen how to back vessels down, literally pushing the dolphins out of the nets. Then, in 1971, innovative purse seiner Harold Medina designed a special panel for the nets, further reducing dolphin entanglements.

Dolphin ensnarement and destruction began during the 1950s in the Pacific Ocean when a special relationship was observed between dolphins and yellowfin tuna. It was discovered that yellowfins congregate and swim beneath dolphins, probably utilizing them as a warning device for larger predators. The dolphins also benefit from a bow wave effect created by the tuna, making swimming easier. Another theory is that the animals share common prey. Whatever the reason, this relationship has been disastrous for both species. Because the dolphins are easily spotted from the air, whole schools can be netted in a process called encirclement. Generally, two boats pull a net around the school of dolphin, thereby entrapping both surface dolphins and the tuna swimming beneath. The net then is closed around the schools and hauled to decks.

These techniques are remarkably successful in killing both species, and only through conservation efforts have dolphin mortalities diminished. Estimated mortalities on U.S. vessels dropped dramatically, from 368,600 in 1972 to 5,083 in 1990. However, while the U.S. mortalities dropped, foreign vessels still reported too many unnecessary dolphin captures. Compared to 5,083 reported by the United States in 1990, the foreign fishery reported 47,448 animals killed that same year. It was obvious to IATTC that an international group needed to be formed to more closely educate and monitor the foreign fishery in an attempt to bring it into compliance. At a landmark meeting in 1992 between countries of the eastern tropical Pacific, a new group under the auspices of IATTC was created. It was called the International Dolphin Conservation Program (IDCP), and its goal was to establish mortality limits of less than five thousand animals per year by 1999. A quota also was established for individual fishermen. Once that number was reached, the individual was required to cease fishing or pay heavy fines. This provided an economic incentive to release dolphins, thereby further reducing kills. In addition, the program demanded 100 percent observer coverage and the creation of a scientific advisory board to oversee the project.

Other options to further reduce dolphin bycatch during tuna fishing are very limited, and each has significant drawbacks. The two primary alternative (dolphin-safe) methods of harvesting yellowfin tuna are the school set and the log set. The school set method nets free-swimming tuna, unassociated with dolphins. The log set targets tuna swimming under logs and other free-floating debris, an environment that normally does not attract dolphin. As tuna congregate under the flotsam, they are captured.

The school set method focuses on immature tuna, those generally under fifty pounds. IATTC scientists estimate that ten thousand sets of purse seine nets will kill 8 dolphins, 2.4 million immature tuna, 12,220 sharks, 1,440 billfish, and 580 sea turtles, to name only part of the bycatch. However, the school set approach has proven to generate less bycatch than the log set method.

The log set method, as estimated by IATTC scientists for ten thousand sets, would kill 25 dolphins, 130 million small tunas, 139,580 sharks, 6,580 billfish, and 1,020 sea turtles. However, in regard to dolphin survival, the school set/log set technique still is an improvement over the use of standard purse seine methods. Those numbers indicate that for ten thousand sets, using standard purse seine methods around mature yellowfins swimming with dolphins, an estimated 40,000 dolphins, 70,000 mature tunas, 520 billfish, and 100 sea turtles would be killed. No sharks were taken with this method.

In 1992, the International Dolphin Conservation Act became law. Effective as of June 1994, it prohibited the sale, transport, purchase, or shipment in the United States, of any tuna that was not caught in a dolphin-safe manner. In order to identify companies independently monitored and compliant with this law, tuna cans were labeled as "dolphin safe." Consumers, in support of the law, have several options. First, purchase tuna sold only by companies that have pledged to remain dolphin safe. Those recommended dolphin-safe brands include StarKist, Bumble Bee, Chicken of the Sea, Kawasho International (Geisha Brand), and Mitsubishi International Corporation. Second, consumers should monitor their grocery store shelves. Managers should be notified if a consumer finds tuna not labeled as dolphin safe. If the store refuses to comply and remove the merchandise, the situation should be reported to the Earth Island Institute's International Marine Mammal Project (IMMP), a nonprofit watchdog agency for dolphin conservation.

This agency also is studying the effects of encirclement on the animals. Besides the direct effect of mortality, the cruel method of encirclement also subjects dolphins to extreme stress. Being chased, herded, and entangled stresses dolphins just as this predicament would stress a human. Effects such as delayed sexual maturity and reduced pregnancy rates may take an eventual toll on the remaining 20 percent of these stocks. Some researchers suggest that dolphins ultimately can learn to adapt and adjust to these pressures. However, encirclement and its impact have yet to be effectively studied.

Unfortunately, in the past several years the U.S. government has relaxed its stand, bowing to pressure from Mexican tuna millionaires and from the World Trade Organization (WTO) to weaken dolphin-safe standards.

Dolphin populations have been declining as a direct result of four decades of fish exploitation, although it appears that numbers are stabilizing. Estimates of spotted dolphins, which account for the majority of dolphin encirclements, and the eastern spinner dolphins are at only 20 percent of their original abundance, but no one knows what numbers are consistent with potential recovery. Scientists estimate, however, that dolphins will be able to replenish their numbers if dolphin-safe standards are maintained.

<hr />

The night's private dolphin display was impressive, but I wanted them closer. I went to the hold and grabbed a fifty-pound box of frozen bait mackerel, tossing the icy fish out into the sea as fast as I could. The dolphin splashed closer, squealing with delight at their good fortune, and in no time at all I had thrown fifty dollars of bait into the sea. It was nearly midnight, time for my watch, and Cooters came looking for me. He caught me red-handed, with the empty mackerel box and dozens of noisy, excited dolphins swimming around the boat. Fortunately, he agreed to forget the incident and not report the wasted bait to the captain, but I had to wise up; otherwise I could be fed to the sharks. I dropped the bait box and headed for the wheelhouse, keenly aware that my irresponsible behavior would not be tolerated. It was time to get back to work.

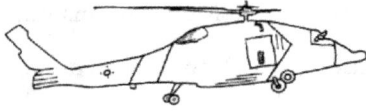

CHAPTER 10
Peril at Sea

We were positioned over the West Wall of the Hudson Canyon, 105 miles northeast of Barnegat Light, when the bright red-and-white, search and rescue helicopter, darted across our bow. Flying at well over 150 miles an hour, the aircraft raced toward the horizon, flying due east as it roared, nose down, against heavy winds. Captain Coyle disappeared into the wheelhouse to turn on the radio and offer assistance. We had completed the day's dismal haul and were free to join him. Moments later, he made radio contact with the source of the Coast Guard emergency. A fellow longliner was in trouble only twelve miles to our east, somewhere near an area called The Elbow in the canyon. The captain of the troubled ship sounded nervous, his voice wavering as he offered a brief explanation of the problem.

We listened intently as the story unfolded, struggling to hear over the whir of the departing chopper. It seems that the crew of the swordboat had pulled gear through the early afternoon. It had been a poor catch, and the last hooks were coming up empty, ready to be stowed. On the last hook, a small, nearly-dead mako shark pup was handlined to the surface and hoisted aboard, then tossed to the stern. The crew already had started cleanup, and a greenhorn high school student was assigned to clean the catch. The captain and remainder of the crew had gone to the galley, but the greenhorn's piercing screams brought them back on deck.

The scene described could have come from a horror movie. The greenhorn was writhing around on deck, shrieking in pain. Blood poured from his right hand, which was engulfed in the jaws of the "dead" mako pup. The crew rushed into action and began beating the little shark. Using gaff poles and baseball bats, the shark finally was killed, its blood splattered over the crew. The damage, however, already was done. The young man's partially severed hand had to be carefully pried from the dead shark's jaws. The crew quickly applied a tourniquet to the arm and attempted to calm the greenhorn while the captain called the Coast Guard station at Barnegat Light, requesting emergency aid. Within minutes, the young man was airlifted to the nearest hospital for emergency treatment by the same helicopter that recently had flown over us. The captain wished the other crew well and we felt relieved, knowing that the young man's life had been saved.[23]

Commercial fishing is regarded as one of the world's most dangerous occupations. As the main source of employment for many communities throughout the eighteenth, nineteenth, and twentieth centuries, one event

---

[23] We later learned that the greenhorn lost his hand, at which point I appreciated my good fortune, keenly aware that I could have been in his place.

at sea could affect an entire town. For example, in 1862, fifteen of the seventy schooners out of Gloucester, Massachusetts were lost while fishing Georges Bank. Seventy women and twice as many children were deprived of a husband, father, and provider. But 1879 proved to be an even worse year for the small New England village when 249 fishermen gave their lives to the sea in that one year alone. Gloucester, in fact, dedicated a statue to her deceased fishermen in 1923 during the seaport's three hundredth anniversary. By the year 2000, the town had created the Wall of Remembrance to honor the more than five thousand Gloucester fishermen who never returned to port. Many other fishing communities have followed suit, including Viking Village on Long Beach Island, which erected a similar monument in 2013, remembering those lives claimed by the sea.

The major threats include storms, sinking, drowning, fires, equipment-related injury, crew disputes, and direct harm from sharks and billfish. Combine that with rough seas, unpredictable weather, and fatigue and you have the ideal environment for catastrophe. U.S. commercial fishermen are thirty times more likely to die on the job than the average American worker. In Italy, the number is twenty-one times the national average, and in Australia it's almost eighteen times, staggering statistics for a job that feeds much of the world's population.

A recent United Nations study reports that fishing at sea claims approximately seventy lives daily. The International Labor Organization estimates that twenty-four thousand fishermen die while fishing every year, and according to the Food and Agriculture Organization, this number may be conservative, given that only a few countries keep accurate records on fatalities.

The northern Atlantic Ocean is known as the graveyard of the Atlantic. Due to the dwindling stocks of fish in local waters, commercial vessels must travel farther offshore and stay at sea for longer periods of time, placing boats and crew at greater risk. Some swordboats from Barnegat Light steam as far northeast as the Grand Banks, some five hundred miles from land.

Stories of rescue at sea are legendary, and in recent years the cooperation between the Coast Guard and working vessels has been the cornerstone of survival. One tale, told by the captain of a Barnegat Light longliner, is worth relating. During the first week of a voyage to the Grand Banks in 1993, while steaming approximately two hundred miles southeast of Nantucket, the panicked mate woke the captain. The VHF radio was transmitting a distress signal from a sinking vessel in the vicinity, and it was learned that the fishermen, all of whom were French, were already in life rafts. The Coast Guard was notified immediately. Minutes later, the longliner approached the raft at the same time as the Coast Guard helicopter. Arriving at the end of a previous mission, the helicopter was unable to continue the rescue due to low fuel supplies and was forced to return to base.

The longliner now assumed full responsibility for both the rescue and well-being of the four French survivors. However, there were now eight, not four, men to feed, and the resources on the small vessel were stretched to the limit. The longliner attempted two more days of fishing, then abandoned the effort and steamed back to Gloucester, Massachusetts, ending a record short trip. Although a financial disaster, all was not lost; the captain received a letter in the mail two months later. He was given a key to the city of a small

French town, the home of the four rescued Frenchmen. Luckily, his heroism was recognized, although this selflessness occurs regularly on the high seas.

But not every rescue ends well for those who attempt to save others. There's the story that occurred after Hurricane Irene. On a trip to Block Canyon, located off Long Island, New York, a Barnegat Light longline captain discovered four men floating on plywood following the sinking of their sixty-foot lobster boat during a storm. One man died, but the remaining men, out of Rockport, Massachusetts, were saved and returned safely to their homes. Although the rescue effort caused a heavy financial loss to the longline crew, the fact is that not one lobster boat crewmen offered a simple thank-you to their rescuers.

Sometimes, rescue is impossible. Sebastian Junger, in his book *The Perfect Storm*, describes one of the most famous commercial sinkings, that of the *Andrea Gail* in 1991. His story explores the life and death realities of commercial fishing and serves to remind millions of Americans of the dangers of this occupation, heretofore only known and understood within fishing communities themselves.

And sinking is only one of the multiple dangers at sea. One such unexpected event occurred in 1997 to another Barnegat Light skipper. While fishing in Block Canyon in five hundred feet of water, he noticed an unusually heavy weight on the gear. As the object reached the surface, the gray mass of a bomb appeared. The crew raced to cut the line, dropping the World War II bomb back into the depths. This time the bomb did not detonate, but the recovery of bombs, depth charges, and unexploded torpedoes is not unusual, and these objects have been known to explode in the hands of an unfortunate crewman.

Drowning at sea also is common. Men are washed overboard during storms, get hooked or tangled in outgoing gear, or simply lose their balance. The most vulnerable time is at night when crewmen are sleeping. A man can fall overboard, his absence unnoticed until the morning when it is too late for rescue. The risk is multiplied by the oilskins that act as an anchor, rapidly filling with water and pulling the unfortunate seaman under.

Survival overboard has been a subject of hot debate for decades. The victim's first and most important goal is to remain afloat by wearing a lifejacket. The second aim is to prevent hypothermia, the chilling of the core temperature due to prolonged exposure to the cold sea. The normal core body temperature is 98 degrees Fahrenheit; unconsciousness occurs at 90 degrees. Cold-water survival depends on several factors including water temperature, body fat, and activity level in the water.

Surprisingly, the active, water-treading victim will cool about 35 percent faster than the still person in a flotation device. Studies indicate that a flotation jacket may extend survival time up to three hours in 50-degree waters. By contrast, the actively moving individual (no flotation device) has only about an hour until the core body temperature drops to 90 degrees.

The advent of survival suits in the 1970s created a tremendous advantage for survival at sea. The suit now is required by all commercial longline vessels; there must be one onboard for each crewmember, at an average cost of eight hundred dollars per suit. The brightly colored yellow or orange suit can be donned in two minutes. Studies show survival rates of nine hours in near freezing water, and much longer in warmer waters. Numerous stories of survival at sea bear this out, and, as technology advances, these numbers should continue to improve.

Crew disagreements are an uncommon source of danger, although rare events have been documented. A particularly strange story involved a U.S. swordboat fishing Canadian waters in the early 1990s. The boat reportedly hired a greenhorn off the "Pick a Dick" board in Gloucester, Massachusetts, before commencing on the two-day trip to Canada. During the outbound voyage, a temporary engine failure caused delays, which disquieted the newest crewmember. The greenhorn reportedly became aggressive and violent at the disruption and threatened the captain's life. After a brief struggle, the captain fatally shot the man in the head. Interestingly enough, the captain eventually was cleared of all charges based on a case for self-defense. This incident illustrates the potential risk of employing new crewmembers, and adds one further element to the dangers at sea.

Another difficulty with hiring unknown crewmembers is the occasional malcontent who, while miles offshore, demands to go home. The transfer of unhappy greenhorns is an entertaining spectacle for seasoned crewmen. The captain first contacts an inbound vessel and upon its arrival the greenhorn is provided a survival suit. He is required to jump overboard and grab the lifeline tossed to him from the inbound ship. The anxious greenhorn then is hauled aboard to face the ridicule of yet another crew.

Another often overlooked safety hazard in the fisheries is equipment failure. Stories of explosions, fires, broken lines, slippery decks, firearm injuries, and damage from various projectiles stretch the imagination. An unusual accident occurred in the Hudson Canyon in June 1979 that permanently affected the lives of at least one of those aboard. The crew was pulling tilefish. Due to a particularly heavy strain on the gear, the hydraulic hauler winch began to overheat. Sparks from the winch touched a nearby gas can and caused a severe explosion. Projectiles struck nearly every crewmember on deck and caused an eye injury to one. Although he was med-evaced by helicopter and given immediate medical attention, the crewman eventually lost his eye.

All that said, aside from the danger of shark bite, swordfish slashing, sinking, drowning, crew violence, and the occasional unexploded bomb, not to mention the increasing danger of having to fish farther from the shoreline in deeper and colder waters, the job is relatively safe compared to, say, a detonation expert.

⋅—⟨⟩⟨⟩⟨⟩—⋅

The *Defiance* rocked on gentle swells, a soothing reminder of our ever-changing environment. We revisited the days' events while stuffing ourselves with filet mignon and baked potatoes, feeling the need to rehearse, at least vocally, the procedures for emergencies and how best to handle oneself in the sharp jaws of a green mako—and, even more dire, man overboard. The captain's pat answer was, "Pray to Poseidon."

I slept poorly that night, dreaming that the *Defiance* was on fire and all aboard had abandoned the burning vessel without life vests for the relative safety of the icy water. I jumped, startled-awake, just as the dorsal fin of a cruising twenty-foot great white began to circle me. My dive watch showed 5:30 a.m., time to start another day with all its potential peril.

⫘⫘⫘ ⫘⫘⫘

There was one layer of protection that couldn't be forgotten. Just at first light, I could hear the booming engines of a Coast Guard HU-25 Guardian Falcon and her big sister, the HC-130 Hercules jet from the safety of the galley. The duo fly daily grid searches looking for boats in distress, fishing violations, and illegal dumping. With a cruising speed of 470 miles an hour and a range of 800 nautical miles, the medium-range Falcon aircraft fly from airbases staggered along the East Coast, combing the Atlantic's 31,530,000 square miles of surface area for trouble. Under the captain's orders to "check the basket," we made sure to leave a laundry basket hanging aloft, within plain view, as a signal that we were fishing. The jet casually inspected us from a high altitude, and then it was gone, leaving us alone again.

During a breakfast of coffee and chocolate doughnuts Cooters recalled stories of the Coast Guard. He explained the "old days" when Coast Guard "boardings" were frequent and the agents carried guns while they searched for drugs and illegal paraphernalia. Sometimes the searches would go on for several hours, and on many occasions confrontations between agents and crew would occur. He also acknowledged that today's Coast Guard is "kinder and gentler." In the 1980s, the Coast Guard changed some of its policies in an attempt to improve interactions at sea; apparently, the new rules helped. Cooters now viewed the Coast Guard as a cooperative agency rather than a governing bully. He readily admitted his respect and admiration for their work.

———⟨⟨⟩⟩———

The U.S. Coast Guard, originally called the Revenue Cutter Service, was created by an act of Congress that became law in August 1790. The service was mandated to supply, and have ready, at least ten boats or cutters to be used for coastal rescue missions and defense. By 1799, the service was reassigned to the newly established U.S. Naval Department, itself based on the traditions and rules of the Royal Navy of Great Britain. However, by 1834, there was one notable exception to this similarity. A Royal Navy tradition, established in the 1700s, allotted each crewman a daily rum ration, dispensed in order to satisfy the crew's need for alcohol, many of whom had been kidnapped from waterfront bars while drunk and pressed into service. But by 1834, the Revenue Cutter Service abolished the daily rum ration, substituting a cash payment of three cents per day, a sum considered high at that time.

By 1847, the Massachusetts Humane Society had established eighteen lifesaving stations along the Massachusetts coastline, creating the U.S. Lifesaving Service, which was intended for shore-based rescue, while the Revenue Cutter Service remained on sea patrol. These two branches merged on January 20, 1915, during the presidency of Woodrow Wilson, and they now are combined under one united front, the U. S. Coast Guard.

One of the most famous mottoes of the Coast Guard is: "You have to go out, but you do not have to come back." The remark originated from the well-known Coast Guard skipper Patrick Etheridge during a rescue effort off Cape Hatteras, North Carolina and was told by CBM Clarence P. Brady, USCG (Ret.), a fellow crewman stationed with Etheridge at the Cape Hatteras Light Saving Station. Brady's account is as follows:

> "A ship was stranded off Cape Hatteras on the Diamond Shoals and one of the life saving crew reported the fact that this ship had run ashore on the dangerous shoals. The old skipper gave the command to man the lifeboat and one of the men shouted out that we might make it out to the wreck but we would never make it back. The old skipper looked around and said, 'The Blue Book says we've got to go out and it doesn't say a damn thing about having to come back.'"

Etheridge's historic statement illustrates the profound danger of the job, thereby acknowledging all those who risk their lives, some of whom have sacrificed their own lives so that others may live.

Today, the U.S. Coast Guard (USCG) is charged with a wide variety of responsibilities including environmental protection, marine safety, and law enforcement. More specifically, these duties include protection of living marine resources, oil spill prevention and cleanup, marine safety (boat safety inspections), public safety education, search and rescue, alien migrant and drug interdiction, and, since 9/11/01, monitoring potential terrorist activities.

Environmental protection has become a central issue of their operations, an effort that has been lauded by numerous environmental groups. In a speech in the year 2000, while addressing the Subcommittee on Oceans and Fisheries, Rear Admiral George Naccara outlined the USCG's increased commitment toward the environment and the role of the First Coast Guard District in protecting the northeastern United States. Under operation Atlantic Venture, cutters and aircraft are assigned to enforce fishery management plans, the Marine Mammal Protection Act (MMPA), and the Endangered Species Act (ESA), which includes over forty species of marine life. As an example of their commitment, in 1999 alone, the USCG expended more than twenty-nine thousand hours of offshore patrolling in order to execute living marine resource regulations.

Enforcing the Magnuson-Stevens Fishery Conservation and Management Act (MSFCMA), the USCG performs on-site, at-sea inspections. Both the Fishery Management Councils (FMC) and the National Marine Fisheries Service (NMFS), working under MSFCMA, have received USCG support, a cooperative effort that certainly will uphold environmental regulations, leading to greater enforcement and protection of the environment.

However, most people know the USCG for its legendary search and rescue operations. Atlantic City Search and Rescue, or SAR, is responsible for protecting boats fishing the deep-water canyons off the New Jersey coastline and possesses a multitude of resources for such offshore operations, including six rotary-wing aircraft (including the HH-65 Dolphin and HH-60 Jayhawk helicopters), three cutters, and numerous surface vessels. The group also supports the Barnegat Light Coast Guard Station, which is restricted to emergencies less than fifty miles offshore.

The USCG utilizes a wide variety of state-of-the-art fixed-wing aircraft, rotary-wing aircraft, ships, and assorted sizes and types of boats in their daily operations, including twenty-six operational Hercules aircraft. SAR currently utilizes two types of fixed-wing aircraft: the aforementioned HU-25 Falcon Jet and the larger HC-130 Hercules long-range surveillance aircraft. The HC-130 behemoth, manufactured by Lockheed, has a wingspan of 132 feet, 4 Allison turboprop engines, and a range of 4,500 nautical miles. At a maximum speed of 380 miles an hour, this aircraft can remain aloft for 14 hours at a time.

While the fixed-wing aircraft play a major role in long-distance surveillance, the rotary-wing aircraft, better known as helicopters, play a key role in rescue operations. SAR's six HH-60 medium-range Jayhawks and the newer HH-65A short-range Dolphin are always on standby, although only two generally are air-ready at any one time. The sixty-five-foot Jayhawk, manufactured by Sikorsky (USA), is stationed out of Cape Cod, Massachusetts; its two General Electric gas turbines make it capable of cruising at an airspeed of over two hundred miles an hour. The smaller Dolphin, manufactured in France by Aérospatiale, is stationed in Cape May, New Jersey. This aircraft measures only 45 feet in length and flies at a slower 190 miles an hour, yet it offers the advantage of improved fuel economy. There are currently thirty-five Jayhawks in active service, compared with eighty Dolphins.

Many types of ships and boats play a significant role in the USCG's activities. A cutter is defined as any vessel sixty-five feet in length or greater, and the USCG utilizes several workboats in this class, including icebreakers, tugs, buoy tenders, and patrol boats. All vessels under sixty-five feet in length are classified as boats, and they usually operate near shore. SAR has three 110-foot cutters currently available, known as island class patrol boats. Named after U.S. islands, these vessels are armed with one .25 mm gun and two .50 caliber machine guns. They are designed like the highly successful British patrol boats and, as their armament suggests, are involved in dangerous law enforcement duties. SAR also maintains smaller patrol boats.

In addition to search and rescue, the USCG has been responsible for organization of inter-vessel cooperation at sea, an idea first discussed after the tragic sinking of the RMS *Titanic* in 1912. Of the 2,200 passengers and crew on board the *Titanic*, 1,495 died in spite of the fact that several ships passed near the site throughout the night. To avoid such future disasters, it was determined that an improved communication system was necessary. But it took forty-six years. Not until July 18, 1958, when the Atlantic Mutual Assistance Vessel Rescue (AMVER) became operational, were registered vessels actually tracked. Information regarding sail plans, ships position, and port of call was logged onto a primitive IBM computer using punch cards and then monitored for vessels in distress.

The system expanded over the next decade and was later renamed the Automated Mutual Assistance Vessel Rescue (AMVER) safety network. It now can create a surface picture of a specific area of ocean showing the relative positions of all participating ships, with the distressed vessel at the center. The search and rescue (SAR) coordinator inputs current ship positions and bearings, and the computerized system then generates a report/printout of ships within the immediate area that could participate in rescue efforts. The computer also will calculate the amount of time required to intercept, based on speed, currents, and weather. The SAR coordinator then directs the ship that is closest to the distressed vessel to respond; this lifesaving communication happens in minutes.

The USCG has continued to improve and promote the AMVER network, and recently it introduced an awards program that has significantly improved participation. Today, over 140 nations, about 40 percent of the world's merchant fleet, participate in AMVER. The network, with an average of twenty-seven hundred ships on the plot every twenty-four hours, has saved over fifteen hundred lives since 1990.

In spite of life-saving communications systems, seagoing vessels also must be equipped to handle their own emergencies. The Commercial Fishing Industry Vessel Safety Act went into effect in 1991 in response to the alarming number of fatalities at sea. It established a checklist of safety items for commercial fishing vessels. The list is expansive yet more common sense than anything else. Specific requirements depend on vessel size and water temperatures. For example, the *Defiance* measured over sixty-five feet long and operated in cold water (by USCG definition, water where the monthly mean temperature is normally 59 degrees Fahrenheit or colder). So she was required to provide: 1) one immersion suit per crewman aboard (they don't make such a device for dogs as of this writing); 2) three-ring life buoys for man overboard; 3) one survival craft; 4) distress signals; 5) fire extinguishers; 6) high water alarms 7) marine sanitation devices; 8) first aid equipment; and, always, 9) the Emergency Position Indicating Radio Beacon (EPIRB).

The EPIRB is a two-foot tube with an antenna and resembles a large flashlight. It is placed within plain sight onboard, usually near the wheelhouse, and must be available on all commercial fishing vessels operating more than three nautical miles from the coastline. This device will float if the vessel sinks, and it sends an automatic distress signal to a satellite. The signal then is interpreted by the rescue coordination center and search-and-rescue units promptly are dispatched.

In addition, the USCG started operation Safe Catch in the winter of 1999. Designed to prevent loss of life and property at sea, it involves voluntary dockside safety exams performed by specialized boarding teams. It also targets high-risk vessels, including those with inadequate safety equipment, those in poor condition, or those with a history of repeated violations. The combination of safety programs has seen marine casualties drop by one half in the First Coast Guard District since 1993.

<hr />

With the Coast Guard plane now gone, we set to work under a sharp sunrise, the reflection obscuring the horizon. I donned sunglasses and searched for the hiflyer but it was Jack who located the start of the gear, only a quarter mile to port. Cooters grabbed a gaff and leaned out over the railing, intent on snaring the object. The crew of the *Defiance* was off to another day of fishing. I smiled, grateful to the Coast Guard, and keenly aware that we were protected from afar. They knew we were here.

CHAPTER 11
Mako Mania

The Northwest Atlantic commercial fisheries primarily utilize two species of large pelagic shark: the porbeagle (*Lamna nasus*), a cousin of the mako, and the larger shortfin mako shark (*Isurus oxyrinchus*), also called bonito shark, mackerel shark, and paloma. A high-speed predator that ranges throughout the warm and temperate oceans, the mako shark's size exceeds twelve feet and one thousand pounds. Increased demand for shark meat has made it a prize target for the commercial swordboat fleets, and, as fishermen have come in close contact with the mako, they have gained great respect for its strength, spirit, and furious struggles when hooked. These qualities make it very difficult to subdue in the water and nearly impossible to control on a rolling deck.

Mako sharks prey on squid and mackerel—and they're the number two enemy of swordfish (man being number one), actively hunting the animals as they bask on the surface of the ocean, making them vulnerable with one bite to the tail. But swords are reported to counter such attacks. In 1988, a Florida sportfishing crew observed a surface battle between a one-thousand-pound mako and a seven-hundred-pound sword. They had just noticed the resting broadbill when they also observed a small wake on the calm surface, created by the tip of the shark's dorsal fin. The shark was closing from two hundred yards behind the smaller fish. The ambush was on. The amazed crew watched as the mako rapidly overtook the resting sword and lunged at the vulnerable tail. The sword, sensing the vibrations, turned and faced its attacker. It dove to meet the shark, outmaneuvering it, driving its sword into the shark's gills. Unable to free its sword, the smaller fish disappeared from sight, taking with it the mortally wounded, impaled mako, as both spiraled into the deep some two thousand feet below, leaving only a blood trail. Locked in this death embrace, this oceanic drama probably took the lives of both animals.

⸻⸻◖⊗⊗◗⸻⸻

The morning was cool, the skies a bright blue. It was day fourteen at sea. Due to the improving weather, the mood was lighter and even Jack seemed to perk up for the first time in nearly a week, discussing his family, his job, and his incessant financial worries. The boat's engines forced us to yell to hear each other, and I shouted at him, trying to express empathy for his hardships. Before we could continue, the blare of the PA system abruptly interrupted the conversation. It was the captain's usual "Let's start the day, ladies" greeting followed by Jimmy Buffett's "Havana Daydreamin'."

It took three hours to pull the set, but all we found was snarled bait and a few fish. Among the disappointing captures, however, was a dead billfish. More waste. We watched the lifeless seven-foot marlin broach

129

the surface. It was devoid of its usual brilliant color and as large and stiff as a surfboard. An eight-year-old male, it had survived from egg to adult with a one-in-several-million chance of ever reaching our gear. We cut the monofilament leader and watched it float away. Because of us, it would complete its life's journey as crab bait. We continued to pull gear, a thankless task with very few fish caught.

With one more mile of ground cable on the main spool and the last of the hiflyers nearly within sight, we talked of lunch while we stripped the empty hooks of bait, watching the small petrels circle each piece of shredded fish that we returned to the sea. We were winding down, beginning to relax, when the captain snapped to attention, our mood interrupted by his body language. The line went taut.

The fish appeared as a blue dot gliding toward our stern in about sixty feet of water. I thought it might be a shark, but, strangely, it offered no resistance against the hydraulic winch of the boat. I wondered if it was saving its strength for a surface encounter. After a sharp curse from the captain, then "small green mako," we scrambled to our stations.

I heard Captain Coyle utter another quick expletive, and then watched as he took control of the situation. It was a confidence-eroding experience, thirty seconds before the encounter, to hear our brave captain say a prayer for help.

I had never encountered a live mako and thought little of it. In fact, I decided to start lunch as the crew brought the fish to boat. But I was wrong. Cooters was first to take action, behaving like a well-trained Marine. Holding the eight-foot gaff, he prepared for the centuries-old battle between man and fish. Jack and I quickly followed his lead, picking up other gaffs, each man pumped full of adrenaline for a battle to the death. We realized that without the usual shotgun aboard, the fish actually would have a sporting chance, despite its now four-to-one odds.

Cooters was first to strike blood. Leaning over the rail, he slipped the gaff into the gills of the 160-pound fish. It hit home, creating a fatal gaping wound to the vulnerable brachial arteries. The fish thrashed fiercely as pumping blood streamed from its head. With each heartbeat, it disappeared momentarily under a crimson murk. I leaned over the port rail, attempting to re-establish visual contact with the fish. Then it surfaced.

Holding the gaff in my sweaty palms, I jammed it into the soft, living flesh around the head. The animal shook violently at the second stick, and Jack buried a third, for security. It took three men, weighing a combined six hundred pounds, to carry the bucking fish through the portside fish door. Once aboard, its fury returned, probably fueled by the lack of available oxygen on deck. Pouring blood from the gills, it rolled and twisted, snapping its jaws at anything within reach, the scene now a frenzied circus with the sounds of the fish pounding the steel deck and men shouting as we twisted away from the snapping jaws. Captain Coyle yelled at the top of his lungs to hold tight to the gaff poles and disappeared for what seemed like hours.

Somehow, during the chaos, I had grabbed the mako's tail. In spite of a cacophony of screams from the now-returned captain and crew to "let go," I remained transfixed in the moment. Still grasping the heavy tail in my hands, the scene unfolded in slow motion. I could see the captain drop to the deck, just in front of the gaping jaws. He carried a board in his left hand and a hacksaw in his right. Although the reason for this equipment was not immediately apparent, I was quick to realize his battle plan and I released the writhing fish. The captain rammed the four-foot timber into the moving jaws as Cooters and Jack climbed aboard the bucking oceanic bronco. The mako was pinned to the deck while, with decisive strokes, the captain sawed at its head like a lumberjack cutting a tree. The shark's agonizing jerks ceased when a final stroke sliced through the spinal cord. It lay still, its head severed. The crew collapsed, covered in the animal's fresh blood. Although it was entirely legal, we had just murdered another living creature. The captain then stood, appearing shaken; he apologized to the crew for forgetting his shotgun, and quickly returned to the wheelhouse, apparently feeling guilty for placing us in this potentially life-threatening encounter.

I struggled to my feet, still incredulous that a fish my size could put up such a titanic struggle against four strong men. The cleanup was like a police crime scene after a gang shootout. The fish's dark blood splattered all across the deck. Its body faced the open fish door, toward the freedom it fought so valiantly to attain. Its head was the trophy we celebrated, the ultimate symbol of conquest. It lay in a large pool of clotting blood, ten feet away from its lifeless body. The scene was one I would never forget for the sheer barbarism of our act. We needed the fish for money, and the mako needed its head. There would be only one winner and one loser in this struggle for survival, and I felt slightly ashamed to relish my status as the ultimate predator.

We talked briefly as we hosed down the bloody deck. Cooters noted the lopsided contest of grown men against one fish. He spoke respectfully of the power and splendor of this mighty animal before tossing the severed head onto the bait table. Then, reaching for the white-handled fillet knife, he carefully sliced the jaws from the head, scraped the remainder of the muscle from the cartilage, and unceremoniously nailed the jaws to the ship's bulkhead to dry. We completed the now-routine cleanup, eager for a break following such a traumatic action.

In an unusual occurrence the captain had fixed lunch, and the only topic during the meal was safety. He praised us for our "cooperation under stress," and we laughed at his intentional understatement. Then he turned his attention to me. He ripped me as a "dumb ass" in grabbing the mako by the tail. He told stories of sharks turning on their backs and biting their tormentor. He then recited the "tiger by the tail" analogy and, seeing my mood darken, moved to our next sore subject. We discussed the poor catch, the prospects for success, and his plan to improve our fortunes.

After lunch, I joined him in the wheelhouse. Although my pride was wounded by his sharp comments, I felt grateful to be alive. Things could have been much worse, and I thought of the young man who had lost his hand to the shark, realizing this could have been me. I stood silent as the captain studied the navigation charts, suddenly hearing a newfound enthusiasm in his voice. Then, without taking his eye from the chart table, he told me to "break out the legs" from the galley freezer. We saved the Alaskan king crab for special occasions, and it looked like this might be one of them.

His excitement became contagious. I scrambled for the freezer, mentally spending my profits on the way down the ladder.

Later that day, to the sound of buttery fingers cracking crab legs, we laughed at the day's events. Cooters calculated that our heroic effort amounted to a four-hundred-dollar fish. After expenses, he continued, we would have earned about thirty dollars per man. We agreed in unison that it seemed crazy to work so hard for a mere thirty dollars. Everything we did on the sea seemed doubly earned. Even downtime was spent working, whether cleaning decks, fixing gear, or planning sets, and we were constantly expending energy. The twenty-hour days were wearing on us, and I noticed we all looked tired. The physical and mental strain was heavy, yet we felt the captain's renewed excitement as we prepared for the evening set.

By 9:30 p.m., we were finished with the set and ready for a game of poker, except for the captain, who returned to the wheelhouse. We cleared the table, sat around on the padded bench, and watched Jack deal the cards, joking that the only way we would make money on this trip was by gambling. Although it generally was prohibited aboard, we kept the stakes very low to avoid monetary disputes. The single light bulb swayed with the swells, casting eerie shadows over the three bedraggled gamblers. Cooters wore the same faded T-shirt he had worn for the past six days and a black bandanna. Jack was dressed in his well-puked-on shirt, and I looked equally pirate-like, sporting a two-week-old beard. But I had never felt better. This moment was an epiphany; I now felt like part of the crew. I had finally passed the last phase of initiation—acceptance.

Cooters smoked his stogie, clouding the room with the odors of cherry tobacco. Waves lapped against the boat, and the almost ever-present Key West sounds of Jimmy Buffett drifted from the wheelhouse, broken by an occasional radio transmission. Maybe it was the quiet. For some reason, the conversation turned serious. Between puffs, Cooters calculated our breakeven point at about three thousand pounds of fish. Since our current catch was only one thousand pounds, each man was technically in debt. Sensing a dampening spirit, he told a few more stories of forty-thousand-pound trips, then ended the game and bid us goodnight. I went on deck, having a rare two hours to relax before my watch.

The sky was clear, the air crisp. With little light pollution so far out at sea, each star shone a magnificent white in the cloudless night sky. As I scanned the surface, I realized we were surrounded by lights on the water. Nearly two hundred smaller sportfishing boats encircled us, reminding me it was a summer weekend. The scene was tranquil, almost trance-like, until a low, guttural growl suddenly broke the peace. Adrenaline rushing, I scanned the deck, catching the stare of my tormentor - Mutt. If only I had drowned him when I had the chance, I thought. But certainly it was not too late to teach him a lesson.

The day had given me courage, strength, and a new resolve. It was now my time, time for the apex planetary life-form to truly dominate. With a quick glance around the stern, I searched the darkness for signs of a weapon. The dog, now closing from about ten feet away, had dropped his head and bared his fangs. His body tensed, every muscle prepared for the attack. He sensed my fear and inched forward.

I spotted a short wooden gaff not four feet away. The handle was just within reach, hanging from the bait table and fortuitously placed. Although my hand shook, I felt both confidence and anger, a lethal combination in any adversary. I was ready to remedy weeks of torment. I waited for him to make the first move. In a flurry of motion, he came at me, his teeth bared, and dove at my leg with an evil growl. This time I was ready for the demon.

With all of my strength, generated by two weeks of suppressed anger, I reached for the heavy oak handle, lowered it, and aimed. The strike only grazed his snout but the beast yelped loudly, then sprinted for the safety of the galley. As I dropped the gaff, I noticed a spot of fresh blood on the deck. I hosed down the area, feeling somewhat relieved. Perhaps my tormentor would finally respect me. It was time for the evening watch.

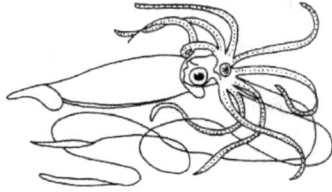

CHAPTER 12
Mystery of the Deep

By 9:00 p.m., we had completed our set on a new location. We were at the Deep Gorge of Hudson Canyon and, according to the captain, two boats fishing this stretch in the past week had captured several unusually large swords. He was excited about our prospects, and his newfound enthusiasm ignited the crew, causing lively galley conversation. The evening passed quickly, and before I knew it, it was nearly midnight. I put aside my copy of *Moby Dick*, leaving Ishmael for my watch.

Jack slouched at the helm, looking tired and sick as usual. He held his right hand that he had cut two days earlier. I studied it, straining to see under the dim light of the radar screen, but caught only a brief glimpse before he pulled away. What I saw looked red; not a good sign, I thought, and probably infected. The fatigue and stress of being at sea weakens the immune system, increasing vulnerability to infection. The fact that Jack remained seasick only worsened matters. His food intake was minimal, and he confided to me that he had lost fifteen pounds in only two weeks onboard. Again, I asked to examine his hand, and he refused, smiling feebly. He asked that I keep the matter between us, not to worry the captain. As he descended the ladder to his rack I watched him move slowly, more like a man of ninety than his actual age of thirty-five. Knowing that he suffered a combination of constant seasickness, fatigue, depression, and now possible infection, I debated about notifying the captain, but after careful consideration decided to wait a couple of days and instead turned my attention to my responsibilities.

By 2:00 a.m. I wanted to sleep. Sitting in the comfortable, worn, wheelhouse chair, I searched the water for the strobe light from the lead hiflyer that signaled the western end of our twenty-five miles of soaking gear. As the boat drifted, separating farther and farther from the longline, my job was to maintain contact with the gear, confirming its position on radar. In the eerie, fluorescent-green glow of the screen, I watched the small blips from the hiflyers as the chain disappeared beyond the radar's ten-mile radius. It was vital to stay awake, and I kept popping my eyes open as soon as I felt my head drop, a pattern familiar to anyone working beyond one's physical stamina. If home watching television the situation would be harmless, but at sea it could carry grave consequences. Collision or losing the gear would be catastrophic. Although beacon buoys send a signal for gear recovery, their range is limited, and twenty-five miles of gear can rapidly vanish into a huge ocean. If the gear is lost, it can be stolen or become entangled with the lines of other boats. It even can become ensnared around the hull of a tanker or freighter. Designed to float sixty feet below the surface (below the draw of an oil tanker), there is no assurance that it is safe.

The situation was becoming grave. I had two hours left in my watch, and my body demanded sleep. I nodded again and then woke with a start. As I looked around, I searched in a panic for the gear. A quick radar check located the set, and I was able to make visual contact with the binoculars. I eased forward on the throttles to pull closer to the gear, hoping not to alert the crew. The *Defiance* vibrated and responded, and I felt her cut into the three-foot swells. Minutes later, we were back on the soaking longline and I felt a rush of relief. Apparently, the captain also was aware of the circumstances and came into the wheelhouse to deliver a mug of steaming-fresh coffee. He smiled and mumbled "good boy" before descending the ladder to his rack. His uncanny omniscience had amazed me from the start of the trip, and I felt empathy for our industrious captain, never able to relax or adequately sleep while at sea.

To stay alert, I forced myself to watch the occasional bright green marks emanating from the fish finder. Apparently, a school of small fish was swimming right under the keel, darting beyond the sonar in seconds. At 2:40 a.m., I noticed the lights of a large ship passing ten miles distant. According to the compass, the ship was steaming southwest, and I followed it on the radar as long as possible, losing contact by 3:00 a.m. Minutes later, I was startled by a strange green blip on the sonar, a massive mark positioned vertically, in six hundred feet of water. It seemed to remain relatively stationary over a five-minute period of time, and then moved vertically toward the surface for about one hundred feet before drifting off the screen. The encounter was brief and remarkable for several reasons: The object was sluggish, impossibly large, and showed significant vertical movement. This was atypical of previous contacts and left me baffled for the remainder of the watch.

The captain was punctual in relieving me at 4:00 a.m. and not particularly interested in hearing about the encounter on the sonar. Rather, he was more concerned with finding fish. He matched LORAN numbers with our position, scanned the radar, and seemed oblivious to my presence. Without making eye contact he suggested that I catch a nap before breakfast. He then returned to his work, sipping coffee, and scratching notes in the ship's logbook. For a brief moment, I considered discussing Jack's health with him, then thought better of the idea as the captain was clearly preoccupied. He needed to find fish; I needed sleep.

The sound was soft and dreamlike as I struggled to wake up. Barely aware of the surroundings, I searched for the source, my eyes rapidly adjusting to the dark. It was Jack, donning his gear. I mumbled something about the time. He understood my gibberish and responded that it was 5:00 a.m. I felt confused, irritated, and certainly not ready to face another twenty-hour day. I staggered to my feet, trying to gain balance and cursing all the while. Jack smiled. We both felt like crap and commiserated over the morning's first cup of coffee. The captain and Cooters already were well into their workday. We sat at the galley table in the dark, listening to them discuss old sea stories. I tried to interest Jack in my sonar encounter of that morning, but he appeared too sick, and tired, to really care.

Up on deck, I noticed Mutt, lying immobile in his corner. His eyes were open, yet he remained still. Without a growl or movement, thinking him dead, I moved closer to the beast. He watched me, without any evidence of aggression. It seemed the previous evening's attitude adjustment had softened him a bit. I smiled, feeling a bit remorseful about the red cut across his snout but satisfied that we had reached a new understanding.

The sky was blue and cloudless over an ocean devoid of sportfishing boats. Then I realized it was Monday morning in the canyon. We were alone, the boats moored in port, and the weekend warriors were back to their mundane day jobs. Without wind, the sun began warming the steel deck. We had a hot day ahead.

The short trip to our set proved fruitless—the gear was nowhere in sight. Following the beacon buoy and radar, we found the line nearly a mile from where we'd anticipated, and that was where the mystery began. We secured the line to our winch and started pulling gear in the typical fashion, but everything was different; something was wrong. The second hiflyer lay horizontally in the swells, the round, bright orange float popped, with a gaping fish-mouth hole in the side. Tangled in the ground cable, the hiflyer was two feet underwater. The huge orange float had split, presumably from the pressure of deepwater submersion. The adjacent gangions were either cut or twisted around the ground line, creating a snarled, tangled mess. Continuing to pull line, we retrieved the mutilated body of a swordfish, only the head remaining. The huge four-hundred-pound animal appeared to have been severed by a single bite. Appearing somewhat surprised, the captain joked that he didn't want to meet the thief that had stolen our swordfish or the cause of the tremendous force required to pull a hiflyer under water. But a third surprise was most startling. We found a single, straightened hook located adjacent to the mangled swordfish. We tried to imagine the monster creature that could cause this amount of havoc and raced to straighten and stow the gear, eager to compare notes.

Sea monsters have long been a topic of heated debate among seafarers, and the captain and crew of the *Defiance* were no exception. Over dinner, we hotly debated the near encounter with this mysterious leviathan, mixing wild speculation and solid fact. The captain remained calm and seemed much less perturbed than the crew as he factually explained the force required to straighten a swordfish hook, noting the manufacturer's specs at five hundred pounds. The crew had their guesses.

Cooters suggested a large shark, and he wasn't completely off the mark. The mighty great white shark (*Carcharodon carcharias*) is an occasional visitor to the Hudson Canyon. They prefer to hunt mammals, especially seals, although they are termed opportunistic feeders and will generally take whatever is available. These huge fish can weigh over three tons and measure twenty-one feet, the largest ever officially documented. It is not unlikely that a twenty-one-foot, three-ton specimen could drag a longline out to sea if it should so choose. And questionable stories abound of fish even twice that length. A famous now-extinct cousin of the great white, the *Carcharodon megalodon,* reached sizes of sixty-plus feet, and many believe these fish still roam the deep.

A large mako was also an option. Fish in the seven-hundred-pound class have been known to become entangled, submerge, and drag gear for miles. Eventually the animal could break free but not before creating a lot of damage. Longline gear is not used for sharks; their sandpaper skin would eventually abrade any monofilament line intended for swordfish, allowing the shark monster his escape.

Jack's guess was a whale, although he really didn't seem to care. Again, this was a legitimate speculation. Whales of all types and sizes inhabit the deep-water canyons along the eastern United States. A large sperm whale could easily damage gear and might even snack on a hooked swordfish. Nearly all species of whale, whether the common finback or the mighty blue, are large enough to drag gear. This seemed to be a very likely scenario, although we had not spotted many whales in the area.

I stuck with my belief that the fabled giant squid (*Architeuthis* sp.)[24] was the probable source and as proof, I related the previous night's strange sonar contact. "Hardly evidence," countered the captain, as he rolled his eyes in mock disgust. But I wasn't convinced. This creature is known to inhabit deep-water canyons and does, on occasion, rise to the surface to hunt.

New England whalers knew of the monster squid. Its arms were found in the stomachs of sperm whales and in their vomit during their final death throes. Although encounters with giant squid have been rare, certain time periods have favored increased experiences. The 1870s, along Newfoundland's coastline, proved just so favorable. More than eighty giants were found stranded and dead. Although the cause still remains a mystery, climatic change, disease, or parasites top the list of likely factors. It is of particular interest that the cold branch of the Labrador Current seems to meander to the Newfoundland coast every ninety years or so, pointing to climatic change as a secondary reason. The giants tend to follow this

---

[24] Kraken, in early literature, was the name for the archetypal sea monster, the giant squid. Its name was derived from early observations that the limbs resembled an uprooted tree; thus the Scandinavian word for "stunted tree" is *kraken*.

cold-water current, resulting in mass stranding when the current runs close to shore. The last stranding occurred in the 1960s, when nine specimens were found between 1961 and 1968. (The next predicted episode will occur in the year 2050, so mark your calendar.)

Interestingly enough, little information is available on the natural history of the giant squid. What is known is that this large and elusive animal lives at depths that are mostly unexplored by man, and it is speculated that many species of squid migrate to the surface at night to feed. One such disturbing event occurred when the *Britannia*, a British troop ship, was sunk on March 25, 1941. Survivors reported that, while drifting in the tropical Atlantic Ocean, a huge squid hauled a man to the deep. Another barely escaped, carrying scars and nightmares to his grave.

The size of these animals has long been a topic for dispute, although there is little doubt that immense specimens are still swimming, undetected. The largest recorded specimen to date was killed at Thimble Tickle, Newfoundland, on November 2, 1878. Although most squid wash up dead, the sickly giant described in the story below put up quite a fight, as documented by the Reverend Moses Harvey:

> On the 2nd day of November last, Stephen Sperring, a fisherman residing in Thimble Tickle ... was out in a boat with two other men: not far from the shore they observed some bulky object, and, supposing it might be part of a wreck, they rowed toward it, and, to their horror, found themselves close to a huge fish, having large glassy eyes,

which was making desperate efforts to escape and churning the water into foam by the motion of its immense arms and tail. It was aground and the tide was ebbing. From the funnel at the back of its head it was ejecting large volumes of water, this being its method of moving backwards.

Finding the monster partially disabled, the fishermen plucked up courage and ventured near enough to throw the grapnel of their boat, the sharp flukes of which, having barbed points, sunk into the soft body. To the grapnel they had attached a stout rope which they had carried ashore and tied to a tree, so as to prevent the fish from going out with the tide ... at length it became exhausted and as the water receded it expired.

The fishermen, alas! knowing no better, proceeded to convert it into dog's meat. It was a splendid specimen—the largest yet taken—the body measuring 20 feet from the beak to the extremity of the tail ... The circumference of the body is not stated, but one of the arms measured 35 feet. This must have been a tentacle.

The Thimble Tickle specimen measured fifty-five feet in length and weighed about two thousand pounds, but calculations based on the circumference of scars left on whales imply that sizes well beyond the Thimble Tickle giant do exist. Sea stories also abound of monster squid encounters, leading many to believe that we have not seen the last of *Architeuthis*.

Another encounter, now well known to both novice and expert teuthologists, occurred during World War II. A man named A. G. Starkey was aboard a British Admiralty trawler stationed off the Maldives in the Indian Ocean and witnessed what is now termed Starkey's Squid. The animal, he reported, measured over 175 feet, making it one of the largest documented squids on record.

Although many stories date back through the centuries, one in particular bears repeating because it relates to a more recent encounter. The experience was related in a letter, written by Mr. Dennis Braun, and submitted to Richard Ellis prior to the completion of his book, *Giant Squid*. It happened near Vieques Island, an area just south of the Puerto Rico Trench, in the deepest part of the Atlantic Ocean, at 27,500 feet.

> In 1969 (about February or March), just prior to my going to Vietnam, I was a 19-year-old Marine Corporal on an amphibious assault training exercise to Vieques Island, near Puerto Rico. We had made an encampment on one of the Island's beaches, stayed for about two weeks and were preparing to return to Morehead City, North Carolina (via the USS. Francis Marion), when I and two of my fellow marines saw something quite extraordinary in the water alongside our ship, which was at anchor.
>
> We had come back aboard ship on the first day of making preparations for re-embarkation, because we had large quantities of equipment to be stowed and there was work to be done in securing this gear in the ship's holds

as it was craned aboard from the shuttling landing craft. During lulls in the activity there was lots of idle time on deck, and the sailors were prepared for this by having their own fishing tackle; so we watched them fish.

The water in that area is very clear, having a light sandy bottom, and with no aquatic growth visible around the ship, you could see straight to the bottom. It was like looking into a swimming pool. The weather was sunny and pleasant and the water very calm. Large yellowtail hung around the ship in schools of 20 or 30, and that's what the sailors were mostly interested in.

We saw sharks and a few other odds and ends, and it was about mid-day on the second or third day that I and two friends were leaning on the port-side railing looking south toward Vieques Island (about half a mile or so distant). Someone from the other side of the ship suddenly came over and announced loudly something like, 'hey, you ought to see the size of the barracuda this guy just hooked!' So naturally just about everybody rushed over to that side to watch the battle. Engrossed in our conversation, and figuring there'd be time anyway, my friends and I stayed behind momentarily—continuing to talk—when I shifted the focus of my vision from the sunlight dancing on the water, to down toward the bottom of the ship. What I saw was astonishing!! There, in full view on the sandy bottom, a huge squid had come to rest! We were amazed as we all three looked on at the thing. Apparently it had just decided to lounge alongside the ship, parallel to it, but still in the sun, and you could clearly see even its eyes as it lay there on the sand; its head to our left, and its tentacles, fanned slightly, extending to our right. The length and girth were truly astounding, and I can only guess as to its actual dimensions, but it was MUCH larger than the one shown laying dead in the TV episode. It just laid quietly very near the ship, being maybe twenty or so feet out from directly below us, and pretty much centered on where we stood (along its length). I would estimate that to scan its entire body length from my position above the water (about 30 ft. up) and by looking straight down, I would have to rotate my eyes at least 30 degrees to the left and 30 degrees to the right.

Its body looked big enough in diameter that I couldn't begin to get my arms around it (not that I'd want to), but maybe halfway. I have little doubt that the thing could have probably taken down a sport-fishing boat, if it chose to, just based on its sheer size and weight! The coloration looked pretty normal for what you'd expect in a squid, brownish, darker than the surrounding sand, but almost like it had taken on some of that lighter color.

There was absolutely no doubt at all as to what it was. The water must have been at least fifty or more feet deep for the ship to be there, yet, in its clarity—and from our place on deck—which was as I would again guess, maybe about thirty feet above the water's surface—we had a very penetrating vantage point. I realize that, based on my assessment of the distance from me, and the angles described above, it was probably at least 100 ft. in length.

It just laid there unmoving and seeming to be looking back up at us as we chattered about it and what to do. I remember suggesting that maybe we should go up to the bridge and tell someone. But no, we reasoned in our naivet,' [sic], they probably already knew it was there because of Sonar and whatever other sensing equipment they might have—our not realizing that sonar wouldn't likely even be switched-on when at anchor. Besides, as young marines, we were all kind of intimidated about approaching officers about anything, much less something like this!

Altogether, I'd say we had watched it for ten or more minutes. Its outline and features were unmistakable, and the time we observed it allowed us to scrutinize and compare our observations pretty much without the benefit of imaginative excess.

I know there are those who would dismiss what I claim to have seen—as drifting seaweed, or the shadows of clouds or the like, but I promise you—if you were to take a highly detailed rubber squid 10 ft. long and sink it in the deep end of a swimming pool—it was just as clear as that. There were no other structural features around it, and Disney couldn't have done a better job of making it real.

149

My dinner companions finally agreed that the most common suspects, in order of decreasing likelihood, would be a very large mako or white shark, a whale, a Navy submarine,[25] or a giant squid. Our irritated captain chimed in, suggesting the probable culprit was an 'angry pink sea serpent'.

⚬⎯⎯⎯⎯(CXXXD)⎯⎯⎯⎯➤

By 7:30 p.m. the gear was dropped overboard. During the entire set Cooters remained unusually animated, telling stories of lost gear, decimated lines, and monster squid. He took great relish in re-telling these tales although we usually doubted his veracity, knowing full-well that he embellished most of the events. The evening was quiet and windless, the seas calm, and the heat of the day held on, drenching us with sweat. This kind of weather, with temperatures in mid-90s and high humidity, left us drained. Like a schooner in the doldrums, we searched for any hint of wind to relieve our suffering. I felt exhausted and retreated to the rack for a two-hour nap before my watch.

Jack was already asleep in his lower bunk. The ten-by-ten-foot closet we slept in felt claustrophobic and stinking hot. The room smelled ripe, much like a low-budget zoo, and as I stretched out in my bunk I caught the disagreeable whiff of sulfur. It seemed that Jack had gas. My patience drained, I headed to the galley where I lay on the table, using a crumpled sweatshirt for a pillow. My misery only worsened when midnight struck. I hadn't slept and my watch was about to start.

---

[25] The presence of Navy submarines operating in the canyons is well known, and the possibility of entanglement is omnipresent, although these events are rare.

CHAPTER 13
Flotsam and Jetsam

The diesels slowed to a halt and then revved into reverse, followed by a series of expletives from the wheelhouse. We raced to the bow to identify the source of the problem. Fifty yards ahead, a wall of debris drifted north to south, stretching as far as the eye could see. As we inched forward to the jetsam, the captain searched for an opening, carefully keeping his distance from visible debris. Floating past the *Defiance* were all shapes and sizes of cardboard boxes, wooden planks, bottles, and plastic bags, all mixed into a potpourri of nearly unimaginable variety. The scene was ugly.

Ocean pollution is a major environmental issue that continues to threaten marine life. It can take many forms, including debris, nonpoint source pollution from various indirect sources, and point source pollution (i.e., chemical plants dumping chemicals into the ocean).

At this time, nonpoint source pollution is the number one problem affecting coastal water quality. Everything from pet waste and lawn fertilizers to toxic household cleaners and oil drippings eventually drains into our waterways and makes its way into the oceans, coating all marine life in a toxic bath. Some of the more visible effects of this pollution include beach closings, algae blooms that sometimes cause huge fish kills, fish consumption advisories, and contaminated drinking water.

The frightening facts are indisputable and a harbinger of things to come. An example of the potential for harm was demonstrated when the Environmental Protection Agency found sixty-seven distinct toxins in urban runoff. Another study found that in New Jersey, over two million pounds (one thousand tons) of pesticides were applied to lawns and farmlands in one year alone, much of this to eventually find its way into the ocean. Perhaps the most shocking fact is that the sewer outflow system of New York/New Jersey flushes billions of gallons of raw sewage into our oceans with every heavy rainfall.

A major type of ocean pollution continues to be human debris. It comes from commercial fishing boats, shipping and transport vessels, recreational fishermen, beachgoers, offshore oil platforms, shore-based solid wastes, plastics companies, and sewage plants during storms. It can include single use plastic bottles and utensils, fishing lines, aluminum beach chairs, straws, fishing nets, bait boxes, light sticks, hard hats, tires, resin pellets, glass bottles, Styrofoam cups, cigarettes, balloons, even toilets and more. It's a frightening statistic; every year, tons of metals, plastics, glass, rubber, and synthetics are dumped into the oceans.

Sadly, the top items removed during beach cleanups tend to be the same from one year to the next, with the major portion of the debris consisting of plastics (60 percent), followed by metal and glass. But the most common, single item found by beach cleanup teams is the cigarette. Cigarette filters not only are an eyesore to area beaches, they also represent a much deeper problem. The filters are made of cellulose acetate, a plastic material that degrades very slowly. In addition, filters contain nicotine, a potent poison, all of which profoundly and adversely affects marine life as these are mistaken for food and ingested by turtles and birds.

Seawater can corrode metal, erode ships, and destroy paper, but it cannot compete with man's ability to pollute. A plastic soda bottle takes about 450 years to biodegrade. Other examples include monofilament fishing line (600 years), disposable diapers (450 years), a tin can (50 years), a Styrofoam cup (50 years), and even the *New York Times*, casually tossed overboard, will take a full 6 weeks to disperse. (The Sunday edition may take even longer!)

The detrimental effect of man's trash on marine life has been elevated to a crime against nature. Seabirds become entangled in active fishing nets, or feed discarded plastic caps to their young, or swallow baited hooks in an attempt to eat the bait. Shearwaters and auklets ingest resin pellets, mistaking them for fish eggs. They also feed discarded trash to their young, resulting in intestinal obstruction and a slow death; birds die from choking, starvation or infection. Sadly, the majestic albatross, a bird that forages across hundreds of miles of open ocean, has been observed feeding its chick a toothbrush, plastic bottle caps and a cigarette lighter, all inadvertently swallowed by the parent as it skimmed the ocean surface for squid and fish eggs.

Five species of endangered or threatened sea turtles inhabit U.S. waters: the loggerhead, green, leatherback, hawksbill, and Kemp's ridley.

These turtles are vulnerable to entanglement and drowning in addition to accidental ingestion of debris. Because the first five years of a turtle's life may be spent drifting with ocean currents, young turtles tend to concentrate in areas where ocean currents converge, and where debris also collects, sometimes becoming gyres or islands of floating waste. Accidental ingestion of plastics often occurs as a case of mistaken identity; single-use plastic bags resemble the turtle's favorite food, the jellyfish. The bags become lodged in the turtle's throat or obstruct digestive passageways and cause the animal to die of starvation.

Whales often become entangled in commercial fishing nets. One report suggested that 56 percent of endangered right whales had scars on their bodies indicating encounters with commercial vessels and fishing gear. The floating nets and lines literally capture the whale, slowing and exhausting the unfortunate animal, leading to death by drowning or starvation. Though rare, the death of a sperm whale found off Seaside Heights, New Jersey in 1996 was proven to be the result of the animal ingesting a garbage bag. Scientists think that free-swimming cetaceans usually have enough sense to avoid devouring man's trash if they have enough natural food available; this is not always the case.

Ghost nets are also a large and unseen pollution problem, due to the six-hundred-year biodegradation time of monofilament line when gear is lost at sea. Gill nets, made of walls of monofilament line, are called ghost nets because they continue to capture and kill marine life long after the nets have been adrift. Fortunately, they tend to ball up and lose some of their killing efficiency with time, but it can take weeks to years for longer nets to collapse, all the while catching and trapping more marine life. In truth, no one really knows how long ghost nets remain a threat in seawater.

The ocean has been used as a dumping ground for waste since the earliest days of sailing. This problem was not acknowledged in the United States until 1899, when it was first recognized as a navigational hazard. In response, the Rivers and Harbors Appropriation Act of 1899 was enacted to prohibit dumping of solid wastes into navigable waters. More recently, the MARPOL treaty[26] was established as an international agreement to prevent ocean pollution from ships. The treaty is divided into several annexes. Annex 1 focuses on oil discharge from ships, and Annex 2 addresses hazardous liquids. Any country that is part of this treaty automatically will adopt the first two annexes. While the remaining annexes are optional, the United States also has adopted Annex V, which deals with plastics and garbage disposal from ships. Violation of this annex carries a fine of up to fifty thousand dollars and imprisonment of up to five years for each offense.

The captain guided the *Defiance* slowly through the slick, searching for an opening, while Cooters moved instinctively to the wheelhouse to grab the binoculars, scanning the water for a clear path through the mess. Finally, his focus shifted to port where he found our escape route, a small patch, fifty yards wide, where the debris parted. Captain Coyle throttled forward, and we inched through the passage, holding our breaths. Finally,

---

[26] The acronym MARPOL is derived from the first three letters in MARine POLlution.

with a desperate lunge and the throttles wide open, the boat burst into pristine open water.

The pressure was on. We had lost fishing time and time is money at sea. The captain maintained radio communications with the boat's owner every two days, and the owner was nervously pressing for more kills. By now, we had close to nine thousand pounds of fish aboard, but we remained financially in the red. There was a growing sense of urgency; we were at sea to produce, meaning to catch fourteen thousand pounds of fish—just to break even. But just breaking even would leave the captain and crew without a paycheck. We had to increase the harvest during the next few days.

Supplies were evaluated and the results were encouraging. We had plenty of meat and vegetables to last another two weeks. Bait and fuel also were in good supply, despite my previous effort to feed the Northwest Atlantic dolphin population. Everyone, however, was concerned about our fresh water supply. We were using more than expected. Potable water means life or death to an offshore vessel, for without it there would be no drinking or cooking water, or worse - no hot coffee.

The weather proved identical to the previous day. Under a tropical high, without any breeze, we baked under a 95-degree midday sun. Everything felt hot, especially the steel deck. We had been at sea for over two weeks, and I noticed that my jeans were becoming stiff with fish scales and dried blood. The odor was offensive, even to the crew, and I searched for options. Explaining my dilemma to Cooters, he taught me the longliner secret to spring-fresh laundry: keel hauling. He tied my Levi's to an old hawser line, connected the line to a portside cleat, and heaved the whole thing into the seas, casually informing me "your laundry should be done in an hour."

The late afternoon sun was low in the sky before the temperature finally dropped to a more manageable eighty degrees. Dinner conversation began with the subject of fishing but quickly changed to more pleasant subjects to lighten the mood. We discussed music and sports and shared a few jokes, avoiding the subject of women to prevent further frustration. The crew finished dinner and disbanded to individual tasks. I scrambled to write notes in my journal, making use of the ten minutes before the start of the evening set.

The routine was familiar; the incredible commotion to the west was not. Two distinct black shapes appeared on the surface, framed by a bright orange sunset. Huge and imposing, the two Navy submarines surfaced, the sailors coming up on deck, probably to take in the same sunset. From a distance of only a mile, the submarines followed our boat for another twenty minutes before their crew disappeared into the massive hulls. Minutes later, the conning towers silently slipped beneath the waves. Although the U.S. Navy frequently patrols the Atlantic submarine canyons, this was to be our only encounter with the most powerful Navy in the world.

The next day was as hot and humid as the previous. By 2:00 p.m. we were preparing for the evening set when we learned that the freshwater pump was broken. Captain Coyle explained that it would have to be repaired before we could resume fishing. Unable to assist with repairs, I searched for the rod/reel in the forward hold and raced to find hooks and leaders. Inside the hold, I found a small white box full of rusted hooks. Tossing a frozen blue mackerel on deck, I quickly baited the line and watched as the fish disappeared beneath the waves.

Twenty minutes later, my efforts were rewarded when a dark fin cruised around our stern. A second fin appeared, and I knew it would only be moments before my bait would get the attention it so richly deserved. The rod suddenly dipped. Sweating beneath a hot, smoking sun I fought to bring the catch to the boat. A four-foot blue dog glided to the stern, but I cut the line close to his jaw, allowing him his freedom. For whatever reason, the game seemed cruel; there were better ways to occupy free time.

By 4:00 p.m. the *Defiance* was back to full operation. Using the PA system, the captain called our attention to starboard, asking "Anyone for a swim?" Gazing onto a flat calm ocean, we could see the tall, brown dorsal fins of hundreds of hammerhead sharks. They slowly glided past us, all going in the same direction, migrating south as they have for nearly a million years, on their way to the Gulf of Mexico. Ten minutes later, they had totally disappeared. Once again I was impressed by the myriad of life forms found in the Northwest Atlantic, in spite of the debris field seen earlier. We never knew what we'd encounter, and day nineteen at sea ended without further incident, the wall of garbage a distant but troubling memory.

CHAPTER 14
Cachalot to Port

By 5:30 the following morning the air finally cooled. The boat rocked on each swell, and the sound of the VHF radio was barely audible against the lapping waves. Jack was deep in sleep and I could overhear the captain and Cooters preparing for the day's work.

Once on deck, I noticed that heavy rain clouds obscured the horizon but, in spite of the weather, I could make out several large black shapes moving northeast about a mile away: a small pod of spouting whales. Not wanting to disturb the captain with such trivial observations, I found my *Tales of Whales* on the chart table and scanned the sections to identify the species. The only obvious clue was the angled blow, visible as a cloud of condensed air at the surface. To my delight, the only whale with this specific pattern was the leviathan, the sperm whale. They seemed indifferent to us and quickly distanced themselves, leaving a new, strangely fishy odor in the air. I sniffed my shirt to identify the source, but the whales seemed to be the source. The whales moved quickly northeast, blowing occasionally and producing a distinctive whooshing sound. Five minutes later, they were gone. It seemed that both man and whale had reached an understanding, developed only in recent years, of peace between the two mammals.

———————◅◆◆◆▻———————

The mighty sperm whale (*Physeter catodon*) has been recorded throughout maritime history and literature as the fiercest and most mystical of the great whales. It is the species of Herman Melville's *Moby Dick*. Sperm whales are the largest of the toothed whales and the largest carnivore on Earth, with males reaching sixty feet and females growing to forty feet.[27] They are found in deep water in all the world's oceans, from the Arctic to tropical seas to the Antarctic. The skin is smooth near the head, becoming indented with long furrowed pits along the body, which probably serve to reduce hydrodynamic drag while swimming.

And they are intelligent. When killer whales and sharks prey on sperm whale calves, the sperm whales attempt to defend their young with a common defensive posture known as the *marguerite flower formation*. The adults arrange themselves like the petals of a daisy, with their heads in, flukes out, thus forming a tight circle around the calf, all ready to use their heavy flukes as battering rams if needed.

The sperm whale dives repeatedly, both day and night, feeding in groups. The animals fan out and submerge in unison, a method advantageous to the

---

[27] The lower jaw of a mature specimen measures ten feet long and contains twenty-five pair of teeth, some of which grow to nearly a foot and a weight of two pounds.

entire group, and can remain submerged for as long as an hour, moving into depths well beyond a mile, constantly in search of fuel for a forty-ton body.[28]

Whales also use echolocation or sonar clicks while hunting. Robert Burton, in his book *The Life and Death of Whales*, states that the "loudness of the Sperm Whales clicks (would be) equivalent to standing six meters behind a jet engine." This intense communication, even in the black abyss, allows them to signal prey conditions to one another over large stretches of water as they search for squid, the main staple of their diet.

Most encounters between sperm whales and giant squid occur unseen in the open ocean; only rarely have these two leviathans have been observed together. A famous and well-known account was recorded by Frank T. Bullen in *Cruise of the Cachalot*, written at the turn of the nineteenth century while the author was fishing the Indian Ocean. Shortly after killing a sperm whale, he described the following:

> At about eleven p.m. I was leaning over the lee rail, gazing steadily at the bright surface of the sea, where the intense radiance of the tropical moon made a broad path like a pavement of burnished silver. Eyes that saw not, mind only confusedly conscious of my surroundings, were mine; but

---

[28] Sperm whales have long been known to feed on squid, generally found in the blackness of the deepest canyons. Their diet also consists of large fish, but exactly how they capture these animals remains a mystery. While many theories abound, some think that the "spermies" utilize bioluminescence to attract prey, thereby luring the food source to the whale and reducing energy expenditure. Sperm whales also hunt sharks, although nearly anything swimming would be considered edible.

suddenly I started to my feet with an exclamation, and stared with all my might at the strangest sight I ever saw. ... A very large sperm whale was locked in deadly conflict with a cuttle-fish, or squid, almost as large as myself, whose interminable tentacles seemed to enlace the whole of his great body. The head of the whale especially seemed a perfect network of writhing arms—naturally, I suppose, for it appeared as if the whale had the tail part of the mollusc in his jaws, and, in a businesslike, methodical way, was sawing through it.

Although it is generally believed that whales pursue squid, several fascinating stories indicate the tables can be turned. The Royal Norwegian Tanker *Brunswick* received attention in the 1930s for her penchant to attract giant squid. She was attacked on three separate occasions, by three different animals, probably because the squid mistook the tanker for an injured whale. The squid paid for their mistakes, however, as they were ground up by the tanker's propellers.

In a rare contemporary account, witnessed in 1965 off the coast of Newfoundland, a battle between the two titans was measured as a draw. A strangled whale was discovered floating on the surface, the squid's tentacles still wrapped around its throat; the whale's stomach later was found to house the squid's severed head.

During the heyday of whaling, there were many reports of encounters between man and sperm whale, although tales of men being devoured by these whales are rare. There is, however, the amazing account that occurred on the whaler *Star of the East* in February 1891. A British seaman named James Bartley apparently survived in the stomach of a sperm whale for sixty hours. Sailing off the Falkland Islands, Bartley's longboat was stove by an alleged eighty-foot bull sperm (twenty feet more than the largest reported in the scientific literature). After a harpoon was planted in the animal's snout, the enraged whale slammed his flukes atop the longboat. Somehow, when the crew was tossed into the sea, Bartley landed directly in the whale's jaws. His horrified shipmates witnessed the event, but returned to their boat, certain that Bartley was beyond help.

The following morning, the crew of the *Star of the East* found a dead eighty-foot sperm whale with a harpoon embedded in his snout. Although the sailors realized it might have been the same whale that swallowed Bartley, no great efforts were made to confirm this possibility. The whale was stripped of its blubber and processed at a leisurely pace over another two days. When the crew finally cut into the stomach, they found Bartley's body, covered in blood and slime, his face purple and twisted in agony, but still alive. The shipmates worked for five hours to resuscitate him. Once recovered, he slipped into a state of crazed panic and had to be tied to a bunk in the captain's cabin for two weeks in order to recover some small degree of sanity. When the ship reached home port, Bartley told his story to a scientific counsel, recounting the details of his sixty hours in the oppressive 104-degree heat and coffin-like blackness of the whale. Being only twenty-one at the time, he actually lived for another eighteen years, all the while displaying on his face and hands the white, bleached scars made by the whale's stomach acid. However, he spent the remainder of his life fearful of the night and unable to look at the sea.

Melville's Moby Dick was actually a sperm whale named Mocha Dick, so named because his home waters were off Mocha Island, Chile. Called the "stout gentleman of the latitudes" because of his huge size, he easily was identified by the eight-foot scar across his huge head. His misadventures reportedly began around 1810, when he first attacked a whaling boat. After Mocha survived his first nasty encounter, he became a lethal rogue, wreaking havoc for nearly sixty years. He was so famous among the Nantucket whalers that their standard greeting was "Any news from Mocha Dick?"

The greatest sea battle between sailors and this whale was recorded in September 1842, some thirty years after Mocha first began his aggression. Following his unprovoked attack on a Japanese lumber ship off the coast of Japan, the three whaling ships, the *Glasgow Criejf*, the *New Bedford Yankee*, and the *English Dudley,* went to the rescue. To the complete astonishment of all, Mocha surfaced a mile distant and proceeded to taunt the crews, posturing in slow, wallowing circles. He faced the three ships and dared they meet his challenge.

The crews of the three ships joined forces and confidently drew lots for who would throw the first harpoon. The small longboats, two from each ship, quickly closed the gap and darted the whale. Mocha played dead for the next five minutes, allowing the boats to close ranks. And then he erupted, furious, crushing the Scottish boat with his twenty-four feet wide flukes. He then turned the English boat into kindling and chewed two men to pulp. Moving through the debris, he proceeded to smash the remaining swimmers before turning back to locate the third longboat. At an estimated swimming speed of thirty miles an hour, he rammed the Japanese boat with such vengeance that the ship rolled bottom up and dropped beneath the waves. The rescuers, overwhelmed and defeated, gathered the survivors and raced for safety into the preferred fury of an approaching storm. The final tally was six men dead, four boats stove, and ninety men defeated by one angry whale.

Mocha Dick's end was not so glorious. Old age had taken a much greater toll on him than man's harpoons. Mocha Dick allegedly was killed by the crew of a Swedish whaler off the coast of Brazil in August 1859. He reportedly measured 110 feet, and at the time of his death appeared weak and crippled. Blind in his right eye, most teeth missing or worn down, his head a mass of scars, he carried nineteen rusting harpoons lodged in his body. He gave up without a fight. But his relatives, with names such as Timour Jack and New Zealand Tom, continued his legacy by terrorizing the high seas for years afterward.

———◦⟨⊗⊗⟩◦———

Back aboard the *Defiance* the whales were long gone, and the crew returned to the daily routine. Under a steady rain and threat of intensifying weather, we hurried to pull the morning set. Despite heavy oilskins, baseball hats and gloves, the cold water still found its way in, soaking our clothes underneath. And to make matters worse, the set was nearly empty, further discouraging the crew. In a three-hour pull that felt like days, the last hiflyer finally came aboard. Looking forward to dry, warm clothes and hot coffee we were quick to store the gear ahead of a fast-approaching storm.

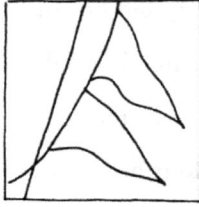

CHAPTER 15
Storm on the Screen

Storms at sea can develop in hours, making an oceangoing ship vulnerable to dangerous wind and waves, sometimes without much warning. The storms form when two air masses of different temperatures come together, creating a front; this is where bad weather begins. Although most storms pass without much concern, a dropping barometer can be a harbinger of death: a full-blown Atlantic hurricane.[29]

Early seafarers learned the warning signs of changing weather by watching the skies, sensing the wind shifts, and remembering the folklore and chants learned in childhood. Such well-repeated verses as "red sky at

---

[29] The word hurricane, from the Spanish *huracan*, causes terror even to experienced seaman. Defined as a tropical cyclone with winds greater than seventy-four miles an hour, hurricanes often occur in the western Atlantic when water temperatures have reached at least 80 degrees. The growing tropical storm picks up moisture as it moves over large areas of ocean. As this warm, moist air rises, the barometric pressure drops, the wind speeds intensify, and the storm builds. Hurricanes continue to gain strength over water, but when they hit land, they lose moisture and weaken.

The Atlantic hurricane season officially begins on June 1st and ends on November 30th.

night, shepherd's delight", "red sky at morning, sailors take warning" and "mackerel sky and mare's tails make tall ships carry low sails" were verbal lessons in maritime weather forecasting. Red morning skies followed by red sunsets signified either rain or snow, while high cirrus clouds, shaped like horses' tails, forecast the approach of strong winds.

Then in 1644, an Italian physicist, Evangelista Torricelli, is credited with designing the barometer, a device that became an invaluable tool to mariners. Used to measure air pressure, the barometer enables seafarers to both predict and monitor oncoming storms, since the first sign of stormy weather is a restless barometer.

Modern seafarers continue to use technology to help predict and avoid deadly storms. The National Weather Service radio provides information on wave height, wind strength, barometric pressures, and long-range forecasts and updates the data every few hours. Another sophisticated instrument is the weather fax, which provides a printout of updated weather maps, complete with a detailed display of frontal patterns and even tropical depressions, while still hundreds of miles away.

For us, the weather picture looked grim. The wheelhouse barometer continued to plummet, the winds increased, and the seas began to build throughout the remainder of the morning. When the captain called us to the wheelhouse for an update, he did not appear pleased. Sighing audibly, he ripped the page from the fax and explained our marine weather pattern for the next twenty-four hours, circling a low-pressure zone near the *Defiance* and a developing storm system toward the southwest. We were too far from shore to beat the storm, so we would have to ride it out. The plan called for cancellation of all crew watches, as the captain would be in the wheelhouse until the storm abated. He suggested that the rack would be the best place for the crew to wait it out.

By 1:00 p.m. the clouds began building to the south, the rains had turned heavy, and cold winds whipped against the wheelhouse windows. I remained with the captain while he stood vigil, listening for radio updates. As the *Defiance* pounded, bow first, into heavy swells and the wind-driven rain pelted the windows, I became increasingly worried, holding onto the chart table, straining to keep my balance.

Between radar checks and VHF updates, the captain talked about previous storms, hurricanes, and many near misses at sea, all of which served to further erode my confidence in our safety. Noting my wide-eyed concern, he laughed and made a brief effort at reassurance before returning his focus to the wheel. Brilliant white lightning streaked across the black sky, allowing me to scan for other ships, but there were none. Visibility was reduced to below a mile. Confirming that the captain was okay, I went below to check on my shipmates. I found Cooters in the engine room, wearing earmuffs and looking most determined to repair something mechanical. He gave me a 'thumbs-up' and smiled, and I moved on to find Jack. He was in his bunk, looking as green as he had looked on our first day at sea. Unable to sleep, read, eat, or be of any use to the crew, I returned to the wheelhouse to wait out the storm.

After nearly three hours, the weather continued to intensify. The storm system first moved south, toward the Bahamas Islands, but now it was headed directly north. By 3:00 p.m. the National Weather Service upgraded the southern low pressure mass from a tropical depression to a full-blown hurricane named Arlene. As she stretched her icy fingers to the north, the worsening weather pattern forced yet another captain's meeting. My fears were confirmed; the captain stated the obvious. It was day twenty-two at sea, and we were low on fuel, bait, and patience. In addition, we were battling a growing storm system that meant more fishing days lost. The trip was over. At 4:00 p.m. we voted to end the trip, sans sick crewmember Jack, whose vote was cast in his absence. We were finally going home.

CHAPTER 16
Compass Southwest

We still had problems: another six hours of steaming time, an intensifying storm bearing down on us, and a dispirited crew. Captain Coyle checked the LORAN numbers for Barnegat Inlet, pointed the bow into our southwestern compass, and inched the throttles forward. It was just before dark, seventy miles east of Barnegat Inlet, when we spotted the strange buoy. It appeared from the mist, bobbing atop heavy swells. With great skill, the captain angled the *Defiance* toward the oddity. Cooters used the gaff to hook a trailing line, and we pulled the heavy object through the fish door. The three-foot-tall object was fire engine red with strange lettering in a foreign language. The crew lashed the mystery jetsam to the deck and quickly forgot about it. Later, the captain was overhead discussing the buoy by radio, and apparently he was advised by that individual to keep the matter quiet. He never mentioned it again, but we would soon discover the importance of the object.

As usual, the wheelhouse boombox played Jimmy Buffett's "Margaritaville," accompanied by our captain's broken serenade. Although the entire crew liked Jimmy's music, we had heard both tapes for nearly a month now and were sick of them. We were dirty, smelly, tired, and frustrated, but we knew the trip would be mercifully over in another six hours. The captain requested coffee, and I headed for the galley, relieved to be useful. I fought for the ladder, just as a large wave threw me to the floor. Rising to my feet, embarrassed, I noticed Jack lying on the table, eyes closed. Moving carefully against obstinate seas, I called his name. He appeared pale. His eyes were glassy, and his hands shook. He was noticeably sweating despite the 60-degree galley temperature. One of his hands was completely swollen and red, but again, he implored me to keep the matter quiet. He alluded to his joy at returning home and his increased hatred of the sea. For Jack, it had been a very difficult trip. I mentioned my concerns about his health, and he agreed to stop at the nearest hospital once ashore.

The evening skies turned a murky darkness as the storm gained momentum on its approach to the northern edge of The Bahamas. The radio reported tides would be low around midnight, forcing us to wait for higher tides at sunrise to enter Barnegat Inlet. The captain and I talked easily from the comfort of the wheelhouse, both feeling satisfied to be going home in spite of the wait and the weather. At 10:00 p.m., I turned on a boombox radio, finally getting reception from familiar Philadelphia radio station-WMMR. They played AC/DC "Back in Black," and I considered this a positive sign, hopeful that I had actually made some money. It also was a relief to hear something other than Jimmy Buffett.

When I asked about profits, the captain turned down the radio. He was obviously frustrated. The *Defiance*, he explained, had returned to port financially "in the red." His statement failed to register with this college boy. He laughed, noticing my blank look, then attempted to explain the situation in terms I could understand, "Eight percent of nothin'," he paused, and with an embarrassed grin continued to say, "... is nothin'." Waiting for my response, he scribbled the numbers I came to despise. The cost of the trip just to go to sea was over nine thousand dollars, including three thousand dollars in fuel, fifteen hundred dollars in bait, six hundred dollars for light sticks, five hundred dollars for ice, eight hundred dollars in groceries, and fifteen hundred dollars in gear, and our catch amounted to only nine thousand dollars of fish. We had barely broken even, a circumstance that I later would learn was common for our captain. (In fact, around Viking Village, he was known as "Captain No-Catch.") I remained mute, stunned by my misfortune. But Barnegat Inlet was dead ahead. The crew of the *Defiance* was heading home; reality would sink in soon enough.

Since this was our first sight of land in nearly a month, Captain Coyle suggested I crack a Budweiser and share one with the crew. I nearly dropped from shock. Here I was, a college student, deprived of alcohol for an unimaginable period of time, and I was invited to have a beer. Diving into the fish hold, I removed three bottles and rejoined the captain and Cooters in the wheelhouse. We shared watch that night, discussing everything from college plans to politics, not wanting the camaraderie to end. As his life's story unfolded, I discovered that the captain was a man with a long history of seafaring. His father taught him to fish at the young age of ten, and he was helping his father operate boats by eleven. He loved the sea, and, even more, he loved to fish. A divorce at an early age had soured his attitude toward any committed relationship. This and his love of the sea helped him feel comfortable being away from land for long stretches of time. As the archetypal swordboat captain, he was demanding but fair, particularly when it came to safety issues or a poor work ethic. He also had developed a strong bond with the boat and spent much of his time repairing and fine-tuning the *Defiance* for service.

Then, before we realized our position, before we were ready, the Barnegat Inlet bell buoy clanged through the mist, and by sunrise, Old Barney was fully visible. The trip was over. By 7:00 a.m. we had navigated the inlet and were tied at our berth. The captain, until now a talkative man, simply said to me, "You're done, go home, great job." I went inside, packed my gear, and returned to the deck. The boat was empty. After nearly a month at sea, all the crew had gone without so much as a "good-bye." But something unusual and most comical was unfolding in the parking lot.

Under the heavy fog and light rain of Arlene's approach, I watched our captain struggle with a long, round object that appeared to be a frozen broadbill. He held the carcass in a bear hug, partially concealed under his yellow rain gear. Placing the deceased in the passenger seat, he observed proper laws and buckled the fish for safety. I watched as the couple disappeared down the quiet road, later learning this was repayment for a debt long overdue.

I wandered around the empty boat, packed but not ready to leave, still thinking about the past several weeks onboard. Feeling physically exhausted yet mentally invigorated, I couldn't wait to tell my story, but first I had to get home. It was Saturday morning on Long Beach Island. Shore traffic was building, and the race was on to escape the islanders preparing for the storm.

Hurricane Arlene tracked toward Bermuda in a broad trough of low pressure, extending into a deep low in the North Atlantic. I was unaware of her path and too tired to realize just how close we were to meeting. Walking to my car, duffel bag in hand, I received several curious looks from local fishermen. I was unshaved, unwashed, and unkempt, and I couldn't have been more pleased. Nearly a month earlier I had arrived on the docks, a total greenhorn, scared and wary, watching similar fishermen boarding their boats to drink beer and fish, an eight-hour trip at most. Now here I was, just off a commercial longliner from the canyons, feeling tough, dirty, and like one of the "big boys." I now felt more like the mighty Captain Quint than the inquisitive Dr. Matt Hooper.

My VW beetle was right where I had left her. She started with a roar, and I was back on the road for home. After battling forty-five minutes of fatigue, dark skies, and a heavy downpour, I arrived to find my older brother and his friends in the kitchen. They were fascinated by my experience, listening for well over an hour to my tale, but after much questioning they broached the sensitive topic of money. Although I tried to deflect the subject, they pushed for the numbers, and I reluctantly admitted to my continuing poverty. Reality had sunk in; I related the captain's words of "8 percent of nothin' is nothin'." They burst out laughing. I knew the chiding had just begun, but I was too exhausted to care. I showed them the swordfish head, shoved it in the freezer, and literally staggered off to bed, still swaying on my sea legs. Sleeping for fourteen hours, my dreams were flooded with marine monsters, sunsets, and hurricanes. I was totally oblivious of Arlene's heavy rains pounding against the bedroom window.

Several days later, recovered from fatigue and my equilibrium regained, I called the captain for details on the mystery buoy. It seems that several government agents showed up at the boat shortly after we docked and confiscated the unidentified floating object (UFO). They also removed the few pictures we had taken. They stipulated that the discovery of the buoy must remain a secret and left as quietly as they had arrived. The whole episode remains unexplained to this day.

Although the trip was over, I still had a college project to complete. The first section of my independent elective was to create a good version of a swordfish skeleton. To do so, I took the sixty-pound broadbill head back to the lab at Stockton State College (now University) where it was introduced to the dreaded dermestids. These large, mean-spirited African beetles eat flesh from bone and can strip a carcass in days. While the beetles labored over the broadbill's head, I began the tedious job of creating a report from the journal I had kept while at sea. I titled the paper "Capture of the King" (a primitive precursor to *The Crimson Broadbill*, the first edition of *The Swordfish Hunters*).

Within two weeks the beetles had performed their grisly task. Now I had to produce a model from the myriad of small bones left by the beetles—my legacy to the college and the first broadbill to grace the walls of the science lab. Without a diagram to follow, I randomly glued pieces together in a loose fashion. Although far from anatomically correct, I just wanted it done. Positioning the final bone in place, and with a sigh of relief, I couldn't help but think of the favorite and oft-repeated phrase of my college adviser, "Close enough for government work." A week later my college paper was accepted, four college credits awarded, and my experience was finally over—or so I thought.

I spent the last week of summer vacation reliving those weeks at sea. I knew that the experience was valuable on several levels. First and foremost, I had passed initiation. I progressed from a greenhorn college boy into a man, a transition that would serve as a building block for the perseverance I would later need to accomplish future goals. Second, the experience fueled my commitment to write a book, to educate the uninformed about the plight of both commercial longline fishermen and swordfish, for both are a dying breed.

But on a deeply emotional level, I had changed. Though I was "hooked" on the thrill of hunting giant fish over the Atlantic's deepest canyons, the memories of the blood and gore remained fresh. It was difficult to partake in the brutal slaughter of such noble creatures, and the very act of terminating life in a violent manner would not, could not, be forgotten.

Epilogue

Captain Coyle fished swords for a few more seasons, but eventually returned to his original occupation in the auto-body repair business. During a recent phone conversation, he discussed both diminished profits and his lack of interest in hard physical labor as reasons for turning his back on the sea. Foregoing his solitary seafaring ways, he has settled into a meaningful relationship and sounded very happy with his new life.

Mutt survived the trip, despite my best efforts, and later was adopted by Cooters. The dog is no doubt long gone by now, but Cooters is alleged to be somewhere in the Caribbean, enjoying the smooth sounds of Jimmy Buffett echoing over those gin-clear waters.

Jack eventually recovered from his wound and returned to his former work. I'm fairly certain he never set foot on a boat deck again.

Although I made subsequent trips aboard the *Defiance,* none were as thrilling as that maiden voyage. During junior year of college I switched directions, choosing a career in medicine instead of marine biology, but

I continued to dream of the sea. I no longer wanted to hunt giant fish - I wanted to save them.

And the dream would not quit. Once settled in medicine, the idea of publishing my shipboard journal kept returning. It was a story that needed to be shared but I wanted assurance that the science was solid; I needed my work reviewed by marine biologists. And I am pleased to state that *The Crimson Broadbill: Commercial Swordfishing the NW Atlantic,* published in 2002, earned positive peer reviews, and a U. N. Atlas of the Oceans listing as a Related Resource – Pelagic Ecosystems. After the book, I launched SandyHook SeaLife Foundation (SSF) in 2006, a non-profit, educational NGO promoting marine conservation.

What drives me today is the realization that swordfish, and other HMS species at the top of the food chain, are disappearing. And though *Xiphias gladius* may rebound in numbers, it will be many years, *if ever,* before 'granders,' or even five-hundred-pound swordfish, are once again common to the deep canyons of the northwest Atlantic Ocean.

SECTION III

THE APEX PELAGICS

ATLANTIC SWORDFISH
HARPOONING TECHNIQUE

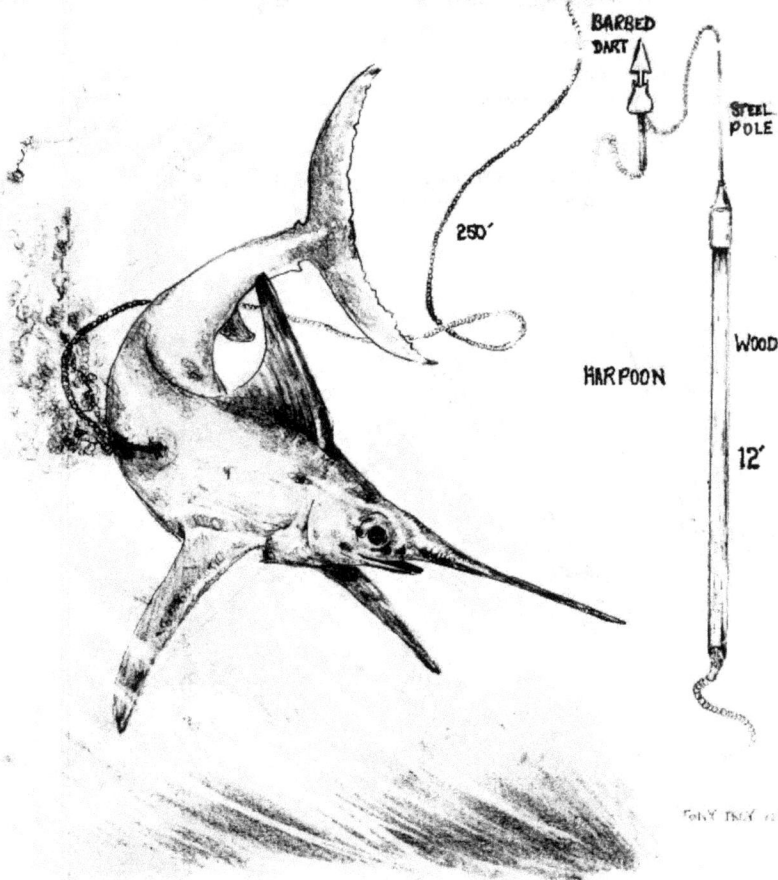

BARBED
DART

STEEL
POLE

250'

WOOD

HARPOON

12'

TONY DALY 2019

Greenstick Gear

1. GREEN STICK
A 35-45 ft fiberglass pole set vertically at or towards
the middle of a boat

2. MAIN LINE
500 to 800 ft of fishing line attached to the green stick
that is towed behind the boat

3. DROPLINE
5 to 10 shorter lines that are attached to the main line
that skips bait across the top of the water

4. SQUID BAIT
Highly colored artificial baits that ship across the water
to mimic a flying fish

5. BIRD
A fish shaped weight found at the end of the main that
holds the line taut while the boat is moving

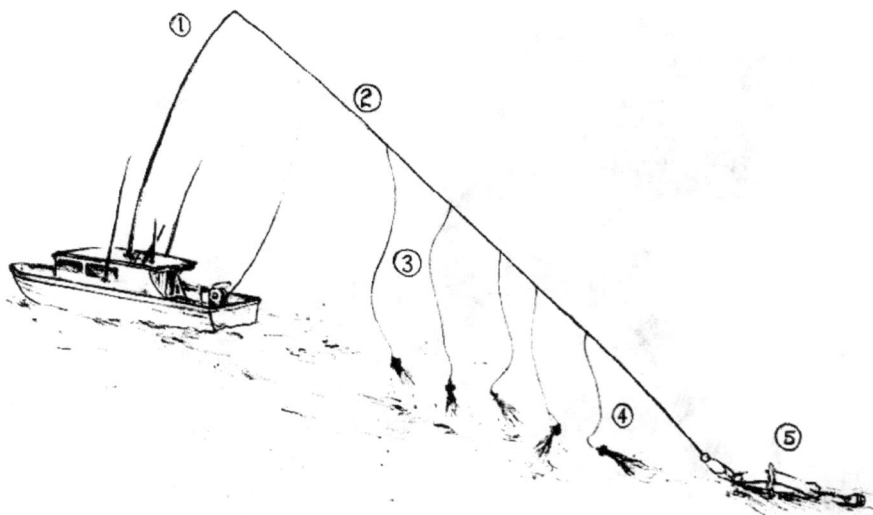

Tony Troy

Buoy Gear

Deep - set buoy gear, illustrated below, is one of the alternative types of gear being explored by researchers and fishermen to catch swordfish and thresher sharks while minimizing harm to other animals.

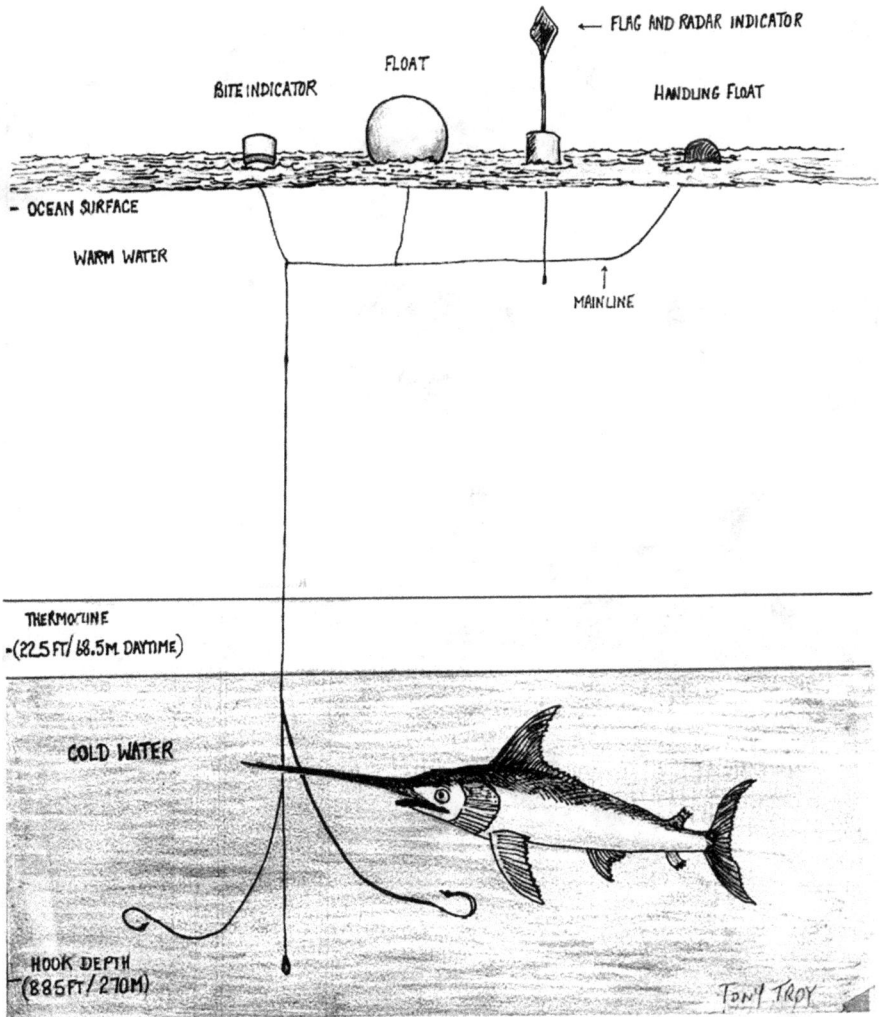

← FLAG AND RADAR INDICATOR

FLOAT

BITE INDICATOR

HANDLING FLOAT

- OCEAN SURFACE

WARM WATER

MAINLINE

THERMOCLINE
- (22.5 FT/ 68.5 M DAYTIME)

COLD WATER

HOOK DEPTH
(885 FT/ 270 M)

TONY TROY

Average Weight in Pounds

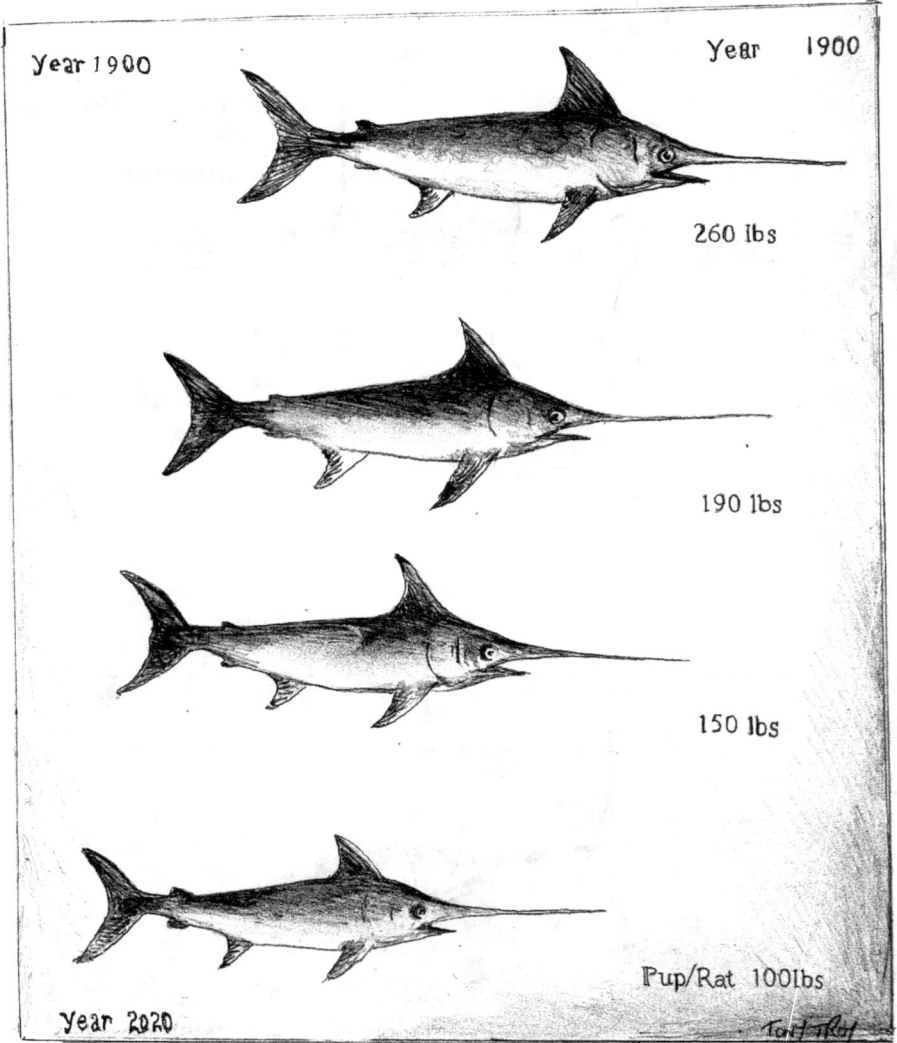

Year 1900

Year 1900

260 lbs

190 lbs

150 lbs

Pup/Rat 100lbs

Year 2020

80% of females caught commercially are less than 150 lbs and are too immature to breed. Virtually all of these undersize "Throwbacks" will die after release.

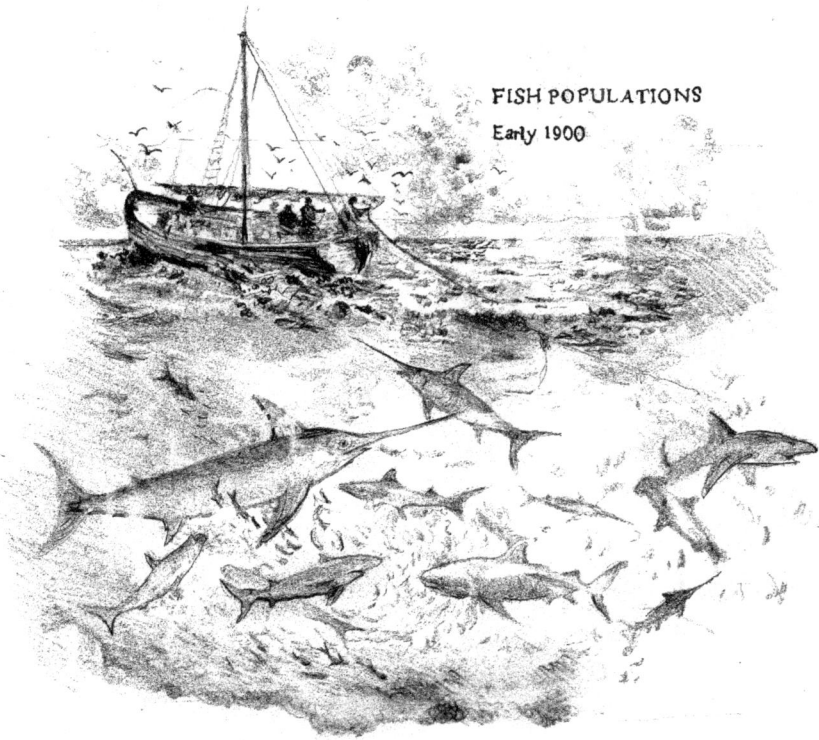

FISH POPULATIONS
Early 1900

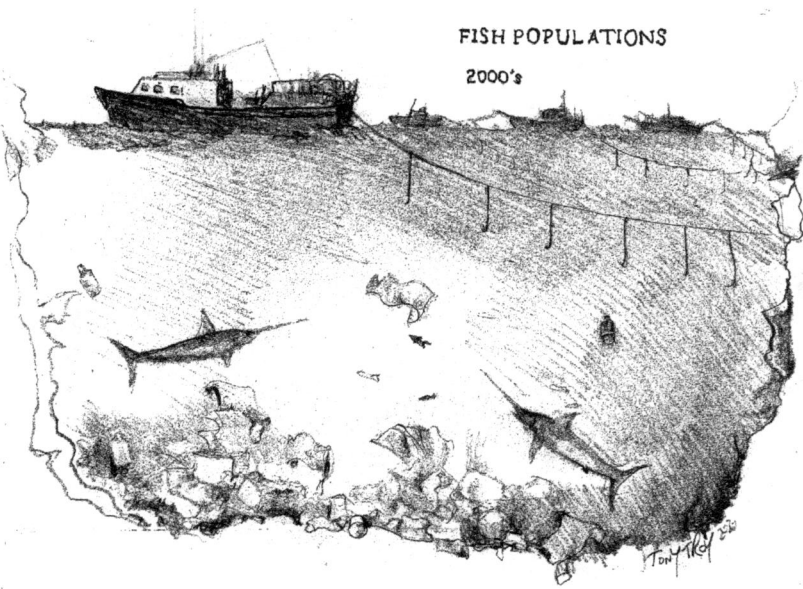

FISH POPULATIONS
2000's

An Overview of the HMS Fisheries in the Atlantic

In Retrospect - prior to 2000

The planet depends on a healthy marine ecosystem, an ocean abundant with life. And that is our challenge, a struggle that reached an acute crisis in the 1990s, when the highly migratory species (HMS) fisheries—comprising three broad groups of large, marine apex predators that include the billfish, tuna, and pelagic sharks— exhibited a freefall in catch rates, largely due to the introduction of longlining in the 1950s.

The Demand

From a historic perspective, Americans tended to follow Adam Smith's theory that he described in *The Wealth of Nations* (1776). Smith believed that individuals somehow made the best decisions for the common good, even when pursuing their own, personal goals. It was as if we were guided by an "invisible hand" that somehow promoted the public interest.

But times have changed. A parallel to our consumer-driven lifestyle was drawn by Garrett Hardin, a recognized author on ethics and ecology. In 1968, Hardin published a treatise on the "tragedy of the commons," the premise of which originated with an Oxford University mathematician, William Lloyd in 1833.

Hardin described how the people of a small village kept all of their cattle in the same pasture. The problem started when some of the villagers felt compelled to increase their holdings and surreptitiously added more cattle to the pasture. Initially, there was plenty of grass to go around, as with the fish supply in the sea. But as the number of cattle increased, the pastureland was eventually overgrazed, the grass supply vanished, and as a result the animals died. Hardin's model applies to today's fisheries. Quite simply, the supply of fish cannot keep up with human demand. The combination of greed and a growing need for seafood has led to "overgrazing" the oceans and the potential outcome could be disastrous to Planet Earth.

When the Japanese began longlining in the Atlantic in the mid-1950s, the method was catastrophic to the HMS population, but it was not the only new challenge. There was also a shift in the top species taken in volume, reflecting actual declines in fish stock. For example, from 1961 to 1973, bluefin tuna accounted for up to 80 percent of the U.S. western Atlantic pelagic catch. In the next quarter century, the numbers declined to less than 15 percent of the total catch, reflecting both a decline in population and newly implemented catch restrictions. The same trend can be documented with the subject of my story, the noble broadbill swordfish. By 1982 it represented 65 percent of the U.S. catch, and two decades later the swordfish accounted for less than 20 (percent).

Another measure of fish population is reflected in the average weight of fish landed. The earliest records indicated swordfish weights between 264 and 411 pounds. Even up to the beginning of the twentieth century, the average whole weight was 300 to 400 pounds, with the largest fish caught at 800 pounds. The record fish documented in the western North Atlantic occurred in 1921, when a robust, dressed, 913-pound trophy fish was brought into Boston. The sword on this old fish was recorded at over four feet long, and the animal's total length was thirteen feet. But by 1963, the average whole weight had dropped to 265 pounds, and the sad fact is that the average weight, just two decades later, was a lowly 84 pounds. This gladiator of the sea was not old enough to reproduce.

A benchmark for assessing the condition of a large pelagic fishery would be to compare the current stock status to the early stock's maximum sustainable yield (MSY). Maximum sustainable yield is defined as the "most fish that can be removed from a population without causing a decrease in a base stock." The fish mortality rate, measured in 1995, was twice the rate associated with maintaining the MSY, while the North Atlantic swordfish biomass (size of a fish population measured in total weight, expressed in metric tons) was estimated to be at only 58 percent of the necessary levels. In simple terms, by the early 2000s the abundance of adult swordfish (age five-plus years) had declined to about one-third since 1985. The depressing scenario is best reflected in the fact that 58 percent of the U.S. swordfish catch were juveniles; they were being caught before they had a chance to breed.

It was clear that by the early 2000s, the classic triad had developed. It included a steady decrease in fish sightings, an increased fishing effort with a diminishing return, and the need for the U.S. swordboat fleets to fish farther from shore in search of prey.

In short, the fishery was declining.

The Free Fall of Other Migratory Species

Other migratory species took the same free fall as the broadbill. For example, the bluefin tuna, once so abundant it was used as cat food, was decimated, with the stock biomass estimated in the early 2000s at only 19 percent of the MSY. The billfish, cousins to the broadbill, were in even graver danger of vanishing. White marlin, prized more highly as a game fish than for food, fell to 15 percent of a level needed for achieving MSY. This population was being commercially fished at seven times the sustainable rate, resembling a graph of the stock market crash of 1929. (As opposed to commercial vessels, recreational fishermen are required to release white marlin alive.)

Pelagic sharks, which represent the third category of highly migratory species, are particularly vulnerable due to a low reproductive capacity, slow growth to sexual maturity, and a long reproductive cycle. This group experienced heavy exploitation beginning in the early 1980s, partly due to the demand for shark fins for Asian markets. The mass-directed fisheries targeted sandbar, blacktip, bignose, tiger, lemon, spinner, and hammerhead sharks, but many other species were killed as well, as bycatch or unwanted capture. The sandbar shark is a good indicator of the shark's precarious position. Studies show that by the early 2000s, this stock had been reduced by 80 to 90 percent in just two decades.

Ever since the popularity of *Jaws* in the 1970s, great white sharks also became targets of sports fishermen, similar to the way big game hunters used to hunt tigers. An archetypal monster, great whites were so aggressively hunted that they became very rare and were considered endangered. A 1997 law addressed the decline of the shark fishery by making the possession of white sharks (and others), dead or alive, strictly illegal.

Additional Damage to the Marine Ecosystem

The clear fact that we were overfishing a declining population was not the only way we were harming the balance of the marine ecosystem. As high-level predators, the migratory species must eat to survive, and their prey was also being overfished. The broadbill feeds on smaller fish such as mackerel and herring, in addition to a heavy diet of squid. While the mackerel stocks remained fairly stable, the Atlantic herring situation was more serious.

The Atlantic herring consists of three major groups in U.S. waters, namely the Gulf of Maine, Georges Bank, and the Nantucket Shoals. The Georges Bank herring stock had collapsed by 1976 due to overfishing by foreign fleets. Given that the fishery also supported a northern broadbill population, Georges Bank was considered fully exploited by the early 2000s.

The northern shortfin squid, a staple in the diet of many top-level predators (including man), had been so overfished in the Atlantic that this once robust fishery collapsed in the 1980s, a fact that created a domino effect in the marine ecosystem. An article printed by the *Seattle Times* (2000), "Fishing Fleets Feast on Squid; But How Long Can the Boom Last?," asserted that the situation on the West Coast was no different. Reflecting an increase in demand for fresh calamari, combined with new markets in China, squid became the leading commercial fish species in California waters. However, many fishermen on the West Coast were uneasy, well aware that the once-productive sardine fishery disintegrated in only thirty-five years. The trickle-down effect of a squid decline could lead to a broadbill collapse as squid is a favored staple in their diet.

The Bycatch Problem

Another major consideration was, and still is, the problem of bycatch, also termed "bykill," referring to the incidental capture of nontargeted, unwanted, or prohibited species or age groups. This problem stems from the nature of the gear used in longlining. Lines stretch from twenty to forty miles or more and contain nearly a thousand baited hooks, all randomly available to any passing animal, whether fish, whale, turtle, bird, or other. Such non-selectivity of longline fishing creates havoc in an already-besieged ecosystem, and the ramifications carry a heavy social, ecologic, and economic price tag.

The numbers caught as bycatch in the years leading up to the crisis in the early 2000s were astounding; nearly one quarter of the global catch was wasted due to bycatch. Putting this into perspective, 44.1 billion pounds of marine life was globally discarded each year, the equivalent of sixty Empire State Buildings. The situation in the pelagic longline fishery was dire—nearly half the catch was returned to the water. Released animals have a poor chance of survival, and greater than 50 percent of these fish already were dead when discarded. Of the remaining living animals cut loose, many did not live very long due to exhaustion and inability to avoid predation. Estimates in the late twentieth century indicated that 80 percent of the billfish discarded in the Gulf of Mexico and off the east coast of Florida by the U.S. longline fishery died in a matter of hours. Examples in terms of numbers, as documented in logbooks, included 50,000 blue sharks discarded in 1993, 40,000 juvenile swordfish discarded in 1996, and over half of these fish did not even survive the trip to the ocean floor. Throughout the 1990s, an estimated 650,000 whales, dolphins, and seals died each year as a result of bycatch. In 1995 alone, an estimated 1,349 extremely rare giant bluefin were discarded, an atrocious waste.

Weight of Responsibility

In the United States, the burden of responsibility fell on several groups to restore this diminished fishery. All United States living marine resources are protected by the National Oceanic and Atmospheric Administration (NOAA). The branch of NOAA directly responsible for the fisheries is the National Marine Fisheries Service (NMFS). This agency regulates fisheries in the three- to two-hundred-nautical-mile U.S. coastal waters, or Federal Exclusive Economic Zone (EEZ), by enforcing the now-highly-disputed Magnuson-Stevens Fishery Conservation and Management Act (MFCMA).

In addition, the offshore fisheries are protected by the Atlantic Tunas Convention Act of 1975 (ATCA), which implements the recommendations of the International Commission for the Conservation of Atlantic Tunas (ICCAT). This management body was created in 1969 to maintain "tunas and tuna-like species" population levels. Although the track record of ICCAT may be disputed, six species under the ICCAT jurisdiction still were being overfished more than three decades after its implementation in 2005, and were well below the MSY.

NMFS' First Attempt

The first attempt to manage the swordfish fisheries began in 1985 when the National Marine Fisheries Service (NMFS) implemented a fisheries management plan based on ICCAT recommendations. Backed by intense research and the blessing of ICCAT, NMFS tightened restrictions in an earnest effort to revamp the fishery. Under a June 1996 ruling, they called for a reduction in the total allowable U.S. catch (TAC) by 359 metric tons to thirty-five hundred tons whole weight. The plan also called for an increase in the minimum size to 33 pounds, required captains to maintain logbook records, and mandated a shorter fishing season. With an allotment of 29 percent to U.S. fisheries toward the world share of North Atlantic swordfish, ICCAT allocated a diminishing total allowable catch to the U.S. share. For example, in 1998 the U.S. share was 3,190 metric tons and 3,103 metric tons in 1999, but now is capped as high as 9,925 metric tons through the year 2021 based on predicted improvement in the fishery. Restricting the quotas certainly will improve the state of the fisheries, but more regulations need to be implemented before restoration can continue.

Wild Oceans (originally - National Coalition for Marine Conservation)

While ICCAT works internationally toward improving stocks, several private, non-governmental groups also are involved. Most notably, in 1973 the organization Wild Oceans: For the future of fishing (originally NCMC) launched the campaign: "Bringing Back the Big Fish" with their NCMC publication "Ocean Roulette: Conserving Swordfish, Sharks, and Other Threatened Pelagic Fish in Long Line-Infested Waters." NCMC/Wild Oceans urged a scheduled phaseout of longline fishing in the U.S. Exclusive Economic Zone: "As currently fished, pelagic longlines are incompatible with effective management of swordfish, billfish and sharks and our national goal of achieving sustainable fisheries."

While calling for the complete phaseout of longliners, NCMC presented thirteen suggestions to protect the remaining highly migratory species. The first four recommendations suggested the prohibition of longlines in spawning and nursery areas where bycatch of juveniles is highest. The closing of these hot spots, termed "time/area closures," would prevent the slaughter of large numbers of juvenile swordfish that congregate in well-known areas of ocean. In August 2000, NMFS enacted historic closures of 133,000 square miles of U.S. coastal waters to commercial longlining; the waters off Florida, the Gulf of Mexico, and the Carolinas were designated as off-limits.

NCMC recommendations five and six called for the installation and use of monitoring systems on individual vessels. Designed to ensure compliance with time/area closures, this method of observation is a computerized tracking device that signaled valuable information directly to satellites, allowing Coast Guard monitoring. Touted as the wave of the future and carrying a four-thousand-dollar price tag, the method hit hard at the already thin wallets of the commercial fisherman. The old method of Coast Guard flyovers and at-sea inspections quickly was giving way to technology.

Recommendations seven and eight suggested limits on gear to a maximum length of ten miles (as opposed to the standard twenty-plus miles), decreasing the number of hooks, and a limit on soak time to six hours (rather than the traditional ten-hour soak time), a concept intended to reduce the number of dead animals by allowing earlier release. (One study showed that marlin and tuna are able to survive up to nine hours on the line, indicating an increase in survival with shorter soak time.)

NCMC recommendations nine through twelve involved gear modifications. This proposal included the use of breakaway gear to be used in areas of high fish concentration, such as the summer migratory routes of the giant bluefin tuna off the mid-Atlantic coastline. Breakaway gear, comprising a lightweight line, would allow more tuna to escape. Such a modification of gear long had been touted as a solution to the longline dilemma. Another example, the use of circle hooks, would enhance survival rates of billfish and tuna since these hooks lodge in the corner of the fish's mouth rather than in the throat. However, the key drawback was that the circle hooks fail with larger specimens that have the strength to straighten the hooks and escape. Breakaway gear modification and its use offered little promise of a solution.

The last recommendation called for counting dead discards against all U.S. quotas for billfish, tuna, and shark, thereby discouraging dead discards, as each animal would count toward the total allowable catch. NCMC also suggested decreasing the number of fishing permits to a maximum of seventy-five by the year 2003 (down from 243) and charging a user fee that would provide for an NMFS observer on each trip. And finally, NCMC recommended continued research on more selective and sustainable methods of fishing for large pelagics.

These thirteen recommendations were proposed in order to encourage a long-term phaseout of commercial longlining but intimated that longlining could be reintroduced if more selective fishing methods were developed and practiced.

The SeaWeb Project

Another nonprofit group closely involved with the swordfish conservation movement was SeaWeb. In partnership with the National Resource Defense Council (NRDC), SeaWeb introduced the "Give Swordfish a Break" campaign in January 1998. It was the first large effort to mobilize seafood consumers and introduce conservation measures to help maintain declining stocks. As a result more than seven hundred professional chefs, as well as airlines, cruise lines, and grocery stores, agreed to remove swordfish from their menus. However, SeaWeb and NRDC ended the formal campaign against swordfish consumption in 2000 with the ICCAT recovery plans in place.

The campaign was well-intended but, since U. S. fishermen were already adhering to the ICCAT quotas (set below the replacement yield level), the boycott needlessly hurt U.S. longline fishermen and their families.

I have worked alongside these diligent, hard-working, and honest people who are part of the U.S. swordboat fleet. They lead a way of life beyond the imagination of most Americans. At sea for over three weeks per month, returning home only to kiss their wives and children good-bye, they live on the cycles of the moon, fishing the first quarter to the last.

Unfortunately, the "Give Swordfish a Break" proved to have only a short-term impact on the consumer, resulting in no appreciable change to the fishery itself.

Foreign Fleets

The Blue Water Fishermen's Association (BWFA), a New Jersey-based organization, represents a great majority of the U.S. longline fleet. During an interview with them, I was reminded of the complexities of the HMS fisheries. Competition for the stock was divided among three main groups: U.S. longliners, U.S. recreational fishermen, and foreign fleets, and statistics provided by the Standing Committee on Research and Statistics (SCRS), the scientists utilized by ICCAT, revealed some startling numbers.

The 2001 SAFE Report for Atlantic Fisheries confirmed that the U.S. vs International catch of North Atlantic swordfish, calendar year 1999, was only thirteen percent of the regional catch and less than 5 percent of the total Atlantic-wide catch of the entire HMS fishery. In other words, approximately 95 percent of the HMS captured in the North Atlantic was taken by foreign fleets. The United States was getting a thin slice of the total pie, and even this small table scrap was further squeezed between U.S. longliners and recreational anglers.

Table 4.1 Calendar Year 1999 U.S. vs International Catch of HMS (mt ww) other than sharks.
Source: NMFS, 2000a

Species	Total International Reported Catch	Region of U.S. Involvement	Total Regional Catch	U.S. Catch	U.S. Percentage of Regional Catch	
Atlantic Sword-fish	40,003 (Atlantic and Mediterranean)	North Atlantic (NA) and South Atlantic (SA)	27,377 (11,914 NA 15,463 SA)	3,087 500 mt discards (2,908 + 494 mt discards NA) (179 + 6 mt discards SA)	13.1% (28.55%NA) (1.20% SA)	8.97% (incl. Med catches)

Section 4: Fishery Data Update   HMS   2001 SAFE Report for Atlantic

The Recreational Fishing Factor

While recreational fishermen charged that longliners were taking all the fish, the numbers published by SCRS told a strikingly different story. In a national report released by SCRS scientists in October 2000, a 1999 comparison of Northwest Atlantic catches of yellowfin tuna purported evidence that the rod/reel (or recreational) take is more than six times that of the commercial longliners. In addition, this only was the reported value, a figure that completely negates the charge that longliners were taking all the fish. And the numbers vary by species.

For example, the bluefin tuna recreational catch in 1999 was over twenty-five times the longline catch for larger bluefin (over 145 centimeters in length). If the scientists are to be believed, the longliners were not taking all the fish. There were certainly exceptions to these numbers, and some species were clearly caught in much greater numbers by commercial boats. The Atlantic broadbill was just such an example, as the rod/reel take is only a sliver of the total American swordfish catch for several reasons. Swordfish are finicky eaters, making them difficult to hook, and they are a challenge to boat due to their famed size and endurance on rod and reel.

Another assertion by the recreational fishermen was that there were too many longliners. In fact, while hundreds of sportfishing boats plied the deepwater canyons every summer, there remained surprisingly far fewer active longliners than expected. The tightly regulated U.S. pelagic longline fishery had allocated fewer than 240 permits for swordfish in the entire Atlantic Ocean since 1999. Even more surprisingly, there were only 60 longliners actively fishing the Atlantic Ocean north of Hatteras, North Carolina, with only 160 active Atlantic pelagic longliners in total. Each of these boats was required to keep a tally for every fish caught, keep a ship's logbook, and carry the VMS computer system to ensure tight compliance with regulations.

By comparison, privately owned recreational anglers were monitored without significant regularity or vigilance and generally seemed to get away with tuna murder. The NMFS does monitor recreational fishermen with occasional dockside intercepts and performs marina-based interviews on a volunteer basis. Telephone interviews also are conducted, but this system only loosely documents the private recreational contribution to the overall assault on the fishery. Additional questions about the compliance of recreational fishermen indicate numerous failures to report catch and the dereliction by some to even obtain tuna fishing permits.

Longline Buyout

A controversial longline buyout plan was introduced to Congress in the fall of 2000. Known as the Breaux/Tauzin Bill, it was revamped for greater impact by New Jersey Congressman, Jim Saxton. Under this proposal (Plan H.R. 4773, the Atlantic Highly Migratory Species Conservation Act of 2000), the Saxton plan required an additional time and area closure of the Mid-Atlantic Bight during peak season for swordfish, tuna, and shark. The bill also allowed the government to purchase all permitted longline vessels up to $400,000 per vessel, depending on the boat's landing history, and allocated a $50,000 buyback of longline permits. To be expected, commercial fishermen saw this bill as an attack on their livelihood. They referred to it as the "Ruin the Livelihood of Commercial Fishermen Act of 2000." Multiple lawsuits were filed and for many reasons, Plan H.R. 4773 never became law.

While marine scientists and conservationists recognize the destructive nature of the method, attempts to phase out longlining have always met staunch resistance from a broad range of stakeholders. It is obvious that the economic and social value of longlining stirs deep emotion. Supporters from commercial and recreational fishermen, seafood suppliers and processors, hospitality/restaurant owners and consumers hold diverse opinions on the best management of this fishing method. But if we continue longlining, our already-diminished wild fisheries will continue on the downward spiral to commercial extinction.

State of the Atlantic HMS – post 2000

In 2017, U. S. fishermen landed approximately 9.9 billion pounds of *all* fish species at U.S. ports at an estimated value of 5.4 billion dollars. In fact, the overall value of all landings between 2016 and 2017 had increased by 2.1 percent with the estimated total annual revenue of the Atlantic HMS fisheries rising to 38.3 million dollars.

But the North Atlantic swordfishery declined. The numbers of swordfish caught by longline in the Northwest Atlantic had dropped from 44,556 in 2013 to 22,332 by 2017. Landings (as measured in metric tons) fell from 1,720.5 in 2013 to 1088.6 in 2015 to 788.1 by 2017. Not surprisingly, annual revenue for swordfish fell by 1.3 million dollars during that same period, with a decrease in both landings weight (boating smaller fish/discards) and ex-vessel price (a combination of the market and the quality of the catch).

Table 5.60   U.S. Catches and Landings (mt ww) of Atlantic Swordfish, by Area and Gear (2011-2015)

Area	Gear	2011	2012	2013	2014	2015
NW Atlantic	Longline*	1,741.8	1,987.0	1,720.5	1,205.6	1,091.0
	Gillnet	0.0	0.0	0.0	0.0	0.0
	Handline	120.4	151.3	104.8	80.9	70.7
	Trawl	17.9	26.8	2.9	5.7	2.9
	Harpoon	0.6	0.3	0.5	0.0	0.0
	Rod and reel**	48.7	64.3	21.7	34.4	45.0
	Unclassified	0.0	0.5	1.6	0.09	0.09
	Unclassified discards	5.8	3.6	0.0	0.0	0.0
Gulf of Mexico	Longline*	363.6	673.3	531.6	300.9	127.4
	Handline	0.5	3.3	0.5	6.7	5.4
	Rod and reel**	4.9	6.3	0.3	1.0	1.0
	Unclassified discards	2.5	6.8	0.0	0.0	-
Caribbean	Longline*	14.2	3.7	20.8	0.2	8.8
	Rod and reel**	0.0	0.2	0.0	0.0	0.0
	Handline	0.0	0.0	0.0	0.3	0.2
	Unclassified discards	0.9	0.0	0.0	0.0	0.0
NC Atlantic	Longline*	451.3	682.6	539.1	309.0	369.0
	Handline	0.0	0.0	.0.0	0.0	0.2
SW Atlantic	Longline*	0.0	0.0	0.06	0.0	0.0
All areas	All gears	2,773.7	3,609.6	2,944.4	1,945.2	1,721.9

* Includes landings and estimated dead discards from scientific observer and logbook sampling programs. ** Rod and reel catches and landings represent estimates of landings and dead discards based on statistical surveys of the U.S. recreational harvesting sector. Source: NMFS 2016.

U.S. Catches and Landings of Atlantic Swordfish by Area and Gear (2011-2015)

However, U. S. consumer demand for seafood continued to rise, creating a growing market for imported products. In fact, more than 90 percent of all seafood sold in the United States is imported. Swordfish imported from all oceans between 2007 and 2014 held steady at approximately 9,442 metric tons, while value of the product increased from $75 to $82 million, with Canada being the leading supplier of North Atlantic swordfish to the United States.

Table 6.23   Imported Swordfish Products (2005-2015)

Year	Fresh (mt)		Frozen (mt)			Total for All Imports	
	Steaks	Other	Fillets	Steaks	Other	(mt)	($ million)
2005	172	6,388	2,957	367	304	10,187	77.17
2006	77	6,830	2,875	351	201	10,334	75.63

Year	Fresh (mt)				Frozen (mt)		Meat			Total for All Imports	
	Fillets*	Steaks	Meat	Other	Fillets	Steaks	> 6.8 kg*	≤ 6.8 kg*	Other	(mt)	($ million)
2007	174	84		5,412	2,520	171	118	737	205	9,422	70.85
2008	96	13		5,658	2,673	170	55	207	88	8,962	68.98
2009	53	10		5,312	1,632	112	96	23	33	7,272	55.85
2010	125	2		5,228	2,077	153	277	45	31	7,939	68.33
2011	74	1		5,060	2,116	139	1,384	471	12	9,258	68.64
2012	13	2	66	5,478	2,013	604	825	43	15	8,993	77.01
2013	31	2	62	6,011	1,394	457	182	4	12	8,093	71.38
2014	31	0	24	7,137	1,575	512	153	<1	32	9,442	82.00
2015	2	162	15	7,751	1,833	578	454	38	56	10,890	87.85

* HTS classification changed as of 2007. NOTE: Imports may be whole weight (ww) or product weight (dw); data are preliminary and subject to change. Source: U.S. Census Bureau.

Imported Swordfish Products (2005-2015)

Table 6.24   U.S. Imports of Swordfish, by Flag of Harvesting Vessel and Area of Origin (2015)

Flag of Harvesting Vessel	Ocean Area of Origin							Total (mt dw)
	Atlantic (mt dw)	North Atlantic (mt dw)	South Atlantic (mt dw)	Pacific (mt dw)	Western Pacific (mt dw)	Indian (mt dw)	Not Provided (mt dw)	
Australia	-	-	-	112.39	122.44	-	-	234.82
Brazil	1.55	0.41	298.51	2.29	-	-	-	302.77
Canada	-	604.09	-	-	-	-	-	604.09
Chile	-	-	-	179.87	-	-	-	179.87
China	-	-	-	2.03	-	-	-	2.03
Costa Rica	-	-	-	615.48	-	-	-	615.48
Ecuador	-	-	-	2,710.06	-	1.50	-	2,711.56
Fiji Islands	-	-	-	7.34	11.48	0.13	-	18.94
French Polynesia	-	-	-	8.49	-	-	-	8.49
Guatemala	-	-	-	0.30	-	-	-	0.30
Indonesia	-	-	-	-	-	63.38	-	63.38
Kingdom of Tonga	-	-	-	0.06	-	-	-	0.06
Malaysia	-	-	-	3.39	-	-	-	3.39
Mauritus	-	-	-	-	-	0.26	-	0.26
Marshall Islands	-	-	-	2.19	-	-	-	2.19
Mexico	-	-	-	291.47	-	-	-	291.47
New Zealand	-	-	-	-	397.10	-	-	397.10
Nicaragua	-	-	-	13.55	-	-	-	13.55
Not Provided	-	-	-	-	-	-	-	0.00
Panama	-	-	-	730.47	-	-	-	730.47
Republic of Maldives	-	-	-	-	-	9.59	-	9.59
Seychelles	-	-	-	-	-	0.54	-	0.54
South Africa	1.95	-	87.09	-	-	44.27	-	133.32
Sri Lanka	-	-	-	-	-	48.87	-	48.87
Trinidad & Tobago	-	5.40	-	-	-	-	-	5.40
Vanuatu	-	-	-	164.63	-	-	-	164.63
Vietnam	-	-	-	101.82	-	-	-	101.82
Total Imports Reported by SDs								6,644.37
Total Imports Reported by U.S. Customs & Border Protection								10,845.86

Source: NMFS Swordfish Statistical Document (SD) Program.

U. S. Imports of Swordfish, by Flag of Harvesting Vessel and Area of Origin (2015)

EXPORTS

Not unexpectedly, U. S. export of swordfish has declined. During the same period (2007–14), exported product dropped from 412 metric tons in 2007 to 156 metric tons only seven years later. Estimates of annual ex-vessel revenue for swordfish indicated steady growth from 2007 to 2013, only to plummet to below 2007 revenues in 2014 (despite the increase in the number of hooks per set from 747 in 2005 to 780 by 2014). By 2015, U.S. caught Atlantic swordfish accounted for less than 8 percent of the total Atlantic catch.

Table 5.2   U.S. vs. Total International Catch of HMS Reported to ICCAT (Calendar Year 2015)

Species	Region		Catch (mt ww)*			U.S. Percentage of Total Catch
			Landed	Discarded Dead	Total	
Atlantic swordfish	N Atlantic	U.S.	1,583	139	1,722	
			10,959	149	11,108	15.5
	S Atlantic		-	-	10,937	0.0
	Total		21,896	149	22,045	7.8
Atlantic bluefin tuna	W Atlantic	U.S.	877	20	897	
			1,816	23	1,839	48.7
	E Atlantic + Med		16,190	11	16,201	0.0
	Total		18,006	34	18,040	5.0
Atlantic bigeye tuna	Atlantic + Med	U.S.	838	-	838	
			79,539	38	79,577	1.1
	Total		78,539	38	79,577	1.1
Atlantic yellowfin tuna	W Atlantic	U.S.	2,076	-	2,076	
			14,701	3	14,704	14.1
	E Atlantic		94,069	137	94,206	0.0
	Total		108,770	140	108,910	1.9
Atlantic albacore tuna	N Atlantic	U.S.	248	-	248	
			25,241	209	25,450	1.0
	S Atlantic + Med		17,852	10	17,862	0.0
	Total		43,093	219	43,312	0.6
Atlantic skipjack tuna	W Atlantic	U.S.	78	-	78	
			19,929	-	19,929	0.4
	E Atlantic + Med		208,652	631	209,283	0.0
	Total		228,581	631	229,212	> 0.1
Atlantic blue marlin	Atlantic + Med	U.S.	9	80	89	
			1,783	81	1,864	4.8
	Total		1,783	81	1,864	4.8
Atlantic white marlin	Atlantic + Med	U.S.	2	8	10	
			457	8	465	2.2
	Total		457	8	465	2.2
Atlantic sailfish	W Atlantic	U.S.	2	6	8	
			886	6	892	0.9
	E Atlantic		1,271	-	1,271	0.0
	Total		2,157	6	2,163	0.4
Blue shark	N Atlantic	U.S.	31	83	114	
			43,583	125	43,708	0.3
	S Atlantic + Med		24,191	128	24,319	0.0
	Total		67,774	253	68,027	0.2
Porbeagle shark	N Atlantic	U.S.	9	35	44	
			21	37	57	77.2

U.S. vs. Total International Catch of HMS Reported to ICCAT (Calendar Year 2015)

Saving a Fishery

Would the initiatives implemented at the end of the 20th century help to improve the conservation status of the North Atlantic swordfish? The answer is yes - to a degree. The temporary ban on the sale and import of swordfish, the area closures of nursery grounds (during breeding seasons), and restrictions on catch limits somewhat improved the western Atlantic stock to the point where it is now considered rebuilt. This means that biomass estimates are currently above the target level. Furthermore, the fishery is listed on the IUCN (International Union for Conservation of Nature) Red List of Threatened Species as "Lower Risk" but qualified as *"Population trend: decreasing."* Statistically speaking, the North Atlantic stock appears to be improving for the immediate future.

However, the last official swordfish stock assessment was released by ICCAT's SCRS in 2017. The report indicated that: "The current TAC (total available catch) of 13,700 tons has an 83 percent probability of maintaining the North Atlantic Swordfish stock in a rebuilt condition by 2021 almost maintaining the level of biomass." But the important words here are *probability* and *almost*.

With the next assessment date yet to be announced, and fishermen reporting mostly juveniles in the western Atlantic fishing grounds, the question is whether this rating can establish, and maintain, a viable commercial swordfishery for the future.

Change Catch Methods

In the long term, selective catch methods like harpoon and deep-set buoy gear must replace longline and gillnets to help achieve sustainability.

The National Marine Fisheries Service currently mandates that commercially caught swordfish measure at least forty-seven inches (with head naturally attached) from the tip of the lower jaw to the tip of the tail. Otherwise, the fish cannot be sold. A fish of this size is not old enough to reproduce, but Atlantic fishermen report that most of the swordfish taken by longline, a multispecies non-targeted gear designed for maximum take, are juveniles that end up as bycatch because they are legally too small to market. This means that while fish are being caught in greater numbers, and the species appears to be rebounding, the Northwest Atlantic swordfish catch consists mainly of young fish and a few large adults. By switching from longline to more selective gear, namely buoy gear or the old method of harpooning, fishermen would boat only targeted adult fish of commercial value.

Buoy Gear Fishing

In comparison to the surface longline method where hooked fish remain in the water for hours, usually overnight, deep-set buoy gear can be used for daylight fishing. But the most significant difference is that the gear indicates when a fish has been caught. This method decreases mortality; the fish struggles less time on the hook.

The gear comprises one or more flotation devices (and sometimes a bite indicator) strung out along a single main line. Usually no more than two squid-baited hooks are attached, but fishermen are allowed a total of twelve to fifteen sets, depending on boat size and permit. The hooks are dropped to a depth of twelve hundred feet, to the cold water where swordfish feed. The fish are attracted to the bait by a light stick or illuminator. When fish take the bait, the gear at the surface moves out of line or bobs, alerting the fisherman to pull the catch, boat it, or return the fish to the water. When a fish meets the required size, it is quickly bled and iced; non-targeted species are immediately released.

Recent studies indicate that buoy gear fishing reduces discards in the number of fish caught and the method produces better outcomes (than longlining) in terms of quality and profit. Buoy gear fishermen report a 94 percent marketable catch rate. The method takes three hundred–plus fish per thousand hooks as opposed to eight swordfish per thousand hooks on longline. In addition, shorter fishing trips mean a fresher product delivered to market, resulting in a better price.

Greenstick Fishing

Another environmentally friendly alternative to longlining in the Atlantic, greenstick is a generic term that applies to all heavy-duty fiberglass and carbon-fiber graphite sticks. Unlike buoy gear that targets swordfish, this active-surface gear also attracts tuna which reports indicate make up 85 to 90 percent of the greenstick catch.

NOAA Fisheries SAFE Report 2015 defines greenstick gear as "an actively trolled mainline attached to a vessel (via a green stick) and elevated or suspended above the surface of the water with no more than ten hooks, and catch may be retrieved collectively by hand or mechanical means." The vessel tows the five to eight hundred–pound mainline across the surface, causing the artificial squid lures to dip in and out of the water like flying fish, attracting the tuna to the surface. The fish take the bait, and unintended catch is released immediately. Greenstick does not disturb sea floor habitat and is proven to reduce the catch of non-targeted species.

The Japanese introduced the method in the early 1980s in Hawaii, and it has been used in the Atlantic since the mid-1990s to harvest bigeye, northern albacore, and skipjack tuna. But greenstick is an especially valuable method for targeting the western Atlantic yellowfin, the population of which has declined by more than half since the 1970s. However, while U.S. buoy gear-caught swordfish is rated sustainable green, U.S. greenstick-caught tuna has not received a similar rating due to the overfished state of yellowfin and bigeye tuna.

The Harpoon

The most sustainable gear may be the oldest – that of the harpoon, a long spear-like instrument used to capture large marine animals and the method used when the Northwest Atlantic swordfishery began. Once a swordfish is targeted the harpoon is thrown, impaling the fish or securing it with barb or toggling claws. Fishermen then use a rope or chain attached to the projectile to hold and draw the animal to the boat.

Harpooning is a highly selective method. Fishermen can target younger, smaller fish, leaving the mature fish to grow to reproductive age. Or they can choose to select large adult fish while allowing immature swordfish to grow to a reproductive age. Of equal importance is that harpooning eliminates the terrible waste of taking non-targeted, unwanted species.

Fish Farming

While ongoing efforts to sustain wild populations of fish remain as the best long-term solutions, discussions on fish farming are germane to the topic. Though it is not possible to farm-raise swordfish (the high-speed nature of the swordfish prevents prolonged survival in captivity), some studies on tuna have shown promise. Although raising the tuna from egg to adult has not yet produced results, the capture and growing of juvenile animals may have potential.

Worldwide, there are multiple sites where bluefin tuna are being raised. Like domestic cattle being fattened for slaughter, the young fish are encircled in large, floating nets in the open ocean, where tons of herring, sardine, and mackerel are pumped into the holding pens from container boats. The increased fat content doubles their value. Once butchered, the fish are flown directly to Japan to be sold for sushi.

Improve Retention Size Limits

Another aspect of fisheries conservation yet to be addressed is that of retention size limits. Current fisheries regulation requires that swordfish can be captured if over 47" (119 cm) in length from lower jaw to tail but this flies in the face of both common sense and research.

In an evolutionary pattern known as Reverse sexual dimorphism, female swordfish are larger than males. A study from Eastern Australia (Young & Drake 2002) found that 50 percent of females reach sexual maturity at 221 cm in length (Age 9.9 years) and males at 101 cm (age 0.9 years). The study also found that two-thirds of retained females caught on longline were immature. This leads to the question of why only larger (breeding) fish over 47" are retained, while immature (nonbreeding) fish are released. It makes more sense to reverse this restriction so that breeding fish are protected and released, while smaller (immature) fish are captured. In other words, why kill a proven breeder, expecting that a younger fish will live long enough to reproduce? This is yet another option to consider if we are to develop a healthy stock of swordfish and applies to other fish species as well.

Implement Cutting-Edge Technology

Commercial fishermen have utilized technology for years, relying on it for vessel safety, voice communications, to locate fish, to track gear at sea, and to detect weather patterns, and the scope of technology has broadened. The emerging discoveries of the Fourth Industrial Revolution, as presented in 2017 at the World Economic Forum's First United Nations Ocean Conference, are designed to address changing ocean conditions and have the potential to prevent overfishing, reduce marine pollution, monitor habitat, count species, gather research and improve profitability, to name just a few. Artificial intelligence (AI), drones, the internet of things, and robotics are currently utilized to improve ocean conditions and make fisheries more sustainable. Newer applications in biotechnology and block chain, currently in various stages of development, will soon be available. What follows is a simplified description of each, as applied to the wild fisheries.

Artificial Intelligence

The Automatic Identification System (AIS) was designed for maritime safety and must be installed for onboard use. The AIS delivers data that identifies the name, course, and speed of the boats nearest the vessel, determines the proximity to other AIS-equipped vessels, and establishes the amount of time needed to avoid collision. AIS and high-resolution satellite imagery systems also allow countries to monitor vessel movements inside and outside of their territorial waters, identifying any fishing vessels operating in forbidden zones.

The Vessel Monitoring System (VMS) is used to monitor fishing activity from land via the movement of commercial VMS-equipped vessels with respect to restricted fishing areas. A given vessel may have approval to fish in a restricted area, to move through it without fishing, or it may not be allowed in the area at all. Some VMS systems also are programmed to record and transmit catch reports.

Drones

Drones—autonomous or remotely operated tools enabled by AI to explore, monitor, and document under water and/or in the air—are used to monitor the ocean, document fish and other sea life in their natural habitats, and collect data on species that are difficult to access. Drones also fill in gaps between on-site data collection and remote satellite imaging and can make working with large animals less dangerous. For example, surveillance drones can spot illegal fishing in a closed area, perhaps in a Marine Protected Area, which could encompass a vast stretch of open water. Drones also can be used to count species underwater, especially in remote environments, to monitor the health of and identify individuals in a pod of whales by collecting DNA samples from the blow holes, and to find lost fishing gear.

Robotics

Robotics—devices that can be electro-mechanical, biological, or hybrid machines enabled by AI—can monitor fish stocks and catches, track fishing vessels to enforce fishing laws, and even trace supply chains. For example, a project currently is underway to create an autonomous underwater vehicle that resembles a live fish in appearance and behavior. It will allow scientists to observe real fish up close and in their natural environment without disturbing marine life.

It has also been suggested that in the future, marine parks might substitute realistic robotic cetaceans for live captives, thereby releasing healthy dolphins, and other marine species, back to the wild. The parks would remain open to the public as educational resources while serving as true advocates for nature.

Internet of Things

An Internet of things is a network of physical devices, vehicles, and other items embedded with electronics, software, sensors, actuators, and network connectivity that enable these objects to collect and exchange data. One possible application consists of cheap, fast sensors aboard a fishing vessel that are used to document the catch. This cost-cutting technology helps fishermen comply with fishing regulations by tracking numbers and, in some cases, the species caught, and by reducing bycatch. In addition, specific commercial fishing vessels that are required to carry human monitors can be equipped instead with electronic devices.

Blockchain

An expanding open list of records accessible to everyone, blockchains have no central authority. The entries are linked and secured using cryptography, thereby providing proof of compliance to standards at origin and along the entire supply chain. For example, each time a fisherman hauls in a catch it becomes a new "asset" on the blockchain. When the catch is sold, the blockchain ID is sold with it. Because blockchain data is nearly impossible to modify, it is considered reliable. Freshly caught fish can be time-stamped and traced from catch vessel to port-of-landing to market and beyond, even to consumer purchase, providing a reliable record of the entire process.

Biotechnology

Biotechnology research in capture fisheries is newly utilized, but the science has already helped to identify species that are more resilient to ocean changes through a process of "assisted evolution," resulting in healthier wild fish. Biotechnology also enables us to measure the differences between populations, thereby identifying unique species to be monitored and improving the health of each fishery as part of a whole.

Current State of Key Highly Migratory Species (HMS)

Levels of Conservation Status

Extinct		Threatened			Lower Risk		
EX	EW	CR	EN	VU	CD	NT	LC
Extinct	Extinct in Wild	Critically Endangered	Endangered	Vulnerable	Conservation Dependent	Near Threatened	Least Concern

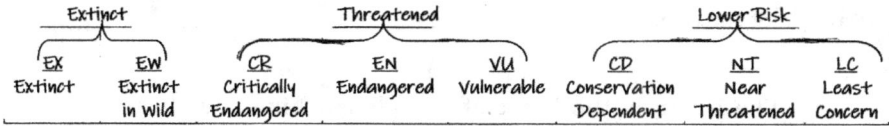

BILLFISH

Swordfish (Xiphias gladius)   NT ⤳ Near Threatened * population decreasing

Blue Marlin (Makaira nigricans)   VU ⤳ Vulnerable

White Marlin (Kajikia albida)   VU ⤳ Vulnerable

TUNA

Bluefin Tuna (Thunnus thynnus)   EN ⤳ Endangered

Big eye Tuna (Thunnus obesus)   VU ⤳ Vulnerable

Yellowfin Tuna (Thunnus albacores) NT ⤳ Near Threatened

SHARKS

White Shark (Carcharodon carcharias) VU ⤳ Vulnerable

Mako Shark (Isurus oxyrinchus)   EN ⤳ Endangered

Porbeagle Shark (Lamna nasus)   CR ⤳ Critically-endangered

INTERNATIONAL UNION FOR CONSERVATION OF NATURE (IUCN)

WWW.IUCNREDLIST.ORG

Why the Ocean Matters

The World Maritime University, a postgraduate university founded by the United Nation's International Maritime Organization, describes the vital role of the ocean:

> The ocean is the heart of our planet. Like your heart pumping blood to every part of your body, the ocean connects people across the Earth, no matter where we live. The ocean regulates the climate, feeds millions of people every year, produces oxygen, is the home to an incredible array of wildlife, provides us with important medicines, and so much more! In order to ensure the health and safety of our communities and future generations, it's imperative that we take the responsibility to care for the ocean as it cares for us.

However, the ocean is in trouble.

Throughout the past 70 years we have damaged this marine ecosystem, acting under the belief that the ocean is impervious to harm and forever capable of sustaining an array of amazing creatures, all thriving in a healthy environment. But science tells us otherwise. The reality is that we have overused the ocean, pouring garbage in and pulling fish out, both at an alarming rate. We have abused the ocean to the point where it is now a microcosm of the most serious threats to the entire planet – a degraded marine environment polluted on many levels – exhibiting food loss, abnormal weather patterns, changes in water quality, vanishing species, damaged habitat and vast shifts in wildlife populations and more.

As valuable marine resources downgrade or disappear, causing yet unknown effects to the entire ecosystem, the human population continues to explode. To quote Stephen Hawking: "We are in danger of destroying ourselves by our greed and stupidity. We cannot remain looking inwards at ourselves on a small and increasingly polluted and overcrowded planet."

The world's population is projected to increase to 11.2 billion people (from 7.5 billion today) by the year 2100, with most of this growth occurring in less-developed countries. By 2050, the United States growth rate could reach 44 percent. Such staggering increases will exceed the Earth's carrying capacity (the maximum population scientists estimate the environment can sustain indefinitely). While the numbers vary from 7 to 15 billion, most agree that we will hit a proverbial "wall" at 10 billion people worldwide. By adding nearly 216,000 human lives per day, the planet will be vastly overpopulated. Factor in poverty, disease, pollution, and climate change and we increase the potential for global conflict.

Will vast populations, including those in third world countries, have access to protein-rich seafood when two-thirds of today's wild fish are already over-exploited? Is it possible to rebuild healthy fisheries in a now-contaminated ocean? How do we reduce, even prevent, ocean pollution, stop wasteful bycatch, and end illegal fishing – all highly-complex issues? There are no easy answers.

Author's Perspective

I can remember returning to the marina after a day of fishing to find dumpsters loaded with dead bluefish, the unwanted catches of the day. At the time, no one questioned the tons of freshly-caught fish left to rot in the sun. We assumed that fish would always be out there for the taking. But technology has enabled us to remove more fish than the ocean can re-produce; the 'tragedy of the commons' is with us. By squandering our marine resources, we destroy ourselves. We can no longer support the large-scale, systematic, and archaic removal of wild fish any more than we can waste captured fish or unintended by-catch.

I suggest ending the cruel, wasteful, and outdated practice of longlining in favor of more sustainable catch methods (harpooning, buoy gear, and greenstick), changing the retention size limits on the highly migratory species, and implementing an ecosystem approach to all fishing. Fishery management plans have focused on a "single species" or on one stock. But future agendas must expand their strategy, take into account the entire food chain, and address the important role of every species within its oceanic community. An example would be the inclusion of forage fish, such as herring, into the overall conservation picture.

And finally, I support U.S. commercial fishermen in these transitions.

What Can You Do?

The ocean's health and the survival of wild fisheries is everyone's responsibility and there are many opportunities to make a difference.

You might start by writing a review of *The Swordfish Hunters.* Or you could study the issues, vote pro-environment, encourage your friends and teach your children to become stewards of the ocean, reduce the use of plastics, walk more/drive less, buy environmentally-friendly products for personal and home use, re-think your seafood choices, volunteer for park and beach cleanups, join and support a marine conservation group – or whatever you choose - but please get involved!

For more information on *Xiphias gladius* refer to NOAA's Technical Report on the Biology of the Swordfish in the following pages. You will find an update on the HMS fisheries in the NMFS SAFE Report 2020 now available online.

Finally, I invite you to visit SandyHook SeaLife Foundation (www.sandy hooksealife.org) and our Facebook page (www.facebook.com/4theocean).

There is much to be done and your choices matter – now more than ever!

For the Ocean,
Dr. Thomas Armbruster

APPENDIX

Glossary of Fishery Terms

Biomass. Biomass refers to the abundance of the stock in units of weight. Sometimes, "biomass" refers to only one part of the stock (spawning biomass, exploitive biomass) but the distinction is not always made.

Bycatch. Catch of species other than the intended target species in a fishing operation. Bycatch either can be discarded or landed.

Commercial. Refers to catch or effort that is commercial in nature, typically using industrial-type vessels and gears.

Discards. Refers to part of the catch that is thrown overboard at sea. Discards may be released either dead or alive. Scientists generally estimate the dead discards as part of the total catch.

Logbook. An official record of a fishing vessel's fishing operations (including location and time of catches, gear configuration, nominal effort used, size samples, etc.). Logbooks are mandatory in some states and are the basis of much of ICCAT's "Task II" data.

Magnuson-Stevens Fishery Conservation and Management Act (MFCMA 1976). Commonly referred to as the Magnuson-Stevens Act, the Magnuson-Stevens Fishery Conservation and Management Act is the legal provision for promoting optimal exploitation of U.S. coastal fisheries.

Overfished. Overfished means that the abundance of the stock is "too low." In many fisheries, the term is used when biomass has been estimated to be below a limit biological reference point that is used as the signpost that defines an "overfished condition." ICCAT has not formally defined when a stock is to be categorized as being overfished, so usage of the term may not always be consistent.

Pelagic. A species that lives in midwater or close to the surface of the open ocean.

Recreational. Refers to catch or effort that is exerted by sportsmen.

Replacement yields. The amount of yield in weight that can be removed from a population of fish and have that stock neither increase nor decline in biomass.

Set. Refers to a fishing operation in which the gear is deployed and retrieved once, usually for purse-seines or longlines.

Stock. In general, a stock is a biological unit of one species forming a group of similar ecological characteristics and, as a unit, is the subject of assessment and management.

Thermocline. A transition zone in the ocean between the upper warm layer and lower cold-water layer. The position of some fishing gears with respect to the thermocline can affect catch rates.

Upwelling. An oceanographic phenomenon that involves the wind-driven movement of dense, cooler, and usually nutrient-rich water towards the ocean surface, replacing the warmer, normally nutrient-depleted surface water.

Yield. The catch in weight. Catch and yield are often used interchangeably. Indicative of amount of production per unit area over a given time or a measure of agricultural production. Yield curve is the relationship between the expected yield and the level of fishing mortality or (sometimes) fishing effort.

Selected References

Armbruster, T. C. "Capture of the King: A Story of New Jersey's Commercial Swordfishery." Outside elective project report, 1986. Richard Stockton State College, NJ.

Avril, Tom. "As swordfish rise, quotas at issue." *Philadelphia Inquirer,* Oct. 2002.

Barnegat Lighthouse State Park System, NJ State Park Commission. Old Barney Tourist Site Information, 2000.

Bathurst, Bella. *The Lighthouse Stevensons.* New York: Harper Collins, 1999, 2, 115.

Belcaro, L. "The Buyout Spin Continues." *Big Game Fishing Journal,* Nov.-Dec., 2000, 6.

Benchley, Peter. *Jaws.* New York: Doubleday, 1974.

Biedeman, N. "US Swordfish Boycott 'Serves' Little Constructive Purpose." Blue Water Fishermen's Association, Barnegat Light, NJ, 1998.

Clean Ocean Action. "Stop Ocean Pollution, Leave Only Footprints in the Sand," 1999.

Clean Ocean Action. "Beach Sweep Results 1999."

Cottingham, D. *Persistent Marine Debris, Challenge and Response to Federal Perspective.* Sea Grant Education Publication, 1988, 6, 7, 13–18.

Cousteau, Jacques. *The Silent World.* New York: Harper & Brothers, 1953.

Dando, M., and M. Burchett. *Sealife: A Complete Guide to the Marine Environment.* Washington, D.C., Smithsonian Institution Press, 1996, 45, 61.

Dietz, T. 1982. *Tales of Whales.* Portland, ME: Guy Gannett, 1982, 27, 59–60, 78–79.

Donofrio, J. "Writing on the Wall: In a Climate Where the Future of Longlining is Grim at Best, Rep. Saxton's Buyout Bill is the Fairest on the Table." *Big Game Fishing Journal,* Nov.-Dec. 2000, 70.

Dore, I. *The New Fresh Seafood Buyers Guide: A Manual for Distributors, Restaurants and Retailers.* New York: Osprey, 1991, 65–68.

Ellis, Richard. *Swordfish: A Biography of the Ocean Gladiator.* Chicago: University of Chicago Press, 2013, 178–79.

Ellis, R. *The Search for the Giant Squid.* New York: Penguin Books, 1998, 78, 84, 86, 90, 92, 136, 245–247.

Federal Requirements for Commercial Fishing: Industry Vessels. USCG brochure. Information current through Oct. 1, 1997.

Fifth Coast Guard District Commercial Fishing Vessel Safety Equipment Requirements. Fifth District Revision VII, 8 August, 1998, USCG.

Folsom, William B., Dale M. Crory, and Karyl Brewster-Geisz. "North America Swordfishing Fleet." In *World Swordfishing: An Analysis of Swordfish Fishing Operations. Past-Present-Future.*" Prepared by the Office of Science and Technology, National Marine Fisheries Service, NOAA, US Department of Commerce, Silver Spring, MD, 1997, 67, 70–71.

Gibson, Charles Dana. *The Broadbill Swordfishery of the Northwest Atlantic.* Camden, ME: Ensign Press, 1998, 3–44, 85–111.

Gordon, J. *Sperm Whales.* Stillwater, MN: Worldlife Library, Voyageur Press, 1998, 7, 11, 14, 22, 29, 37.

Greenlaw, Linda. *The Hungry Ocean: A Swordboat Captain's Journey.* New York: Hyperion, 1999, 259–61.

Hendrickson, R. *The Ocean Almanac.* New York: Doubleday, 1978, 79–80, 97, 131, 409.

Heuvelmans, B. *In the Wake of the Sea Serpents.* New York: Hill and Wang, 1965, 63–65, 68–71, 78.

Hinman, K. *Ocean Roulette: Conserving Swordfish, Sharks, and Other Threatened Pelagic Fish in Longline-Infested Waters*. National Coalition for Marine Conservation, Leesburg, VA, 1998, 7–8, 32, 35, 61, 83, 87–90.

International Coastal Cleanup, 1997 Results. Center for Marine Conservation, 1998, 12.

Kraft, Bayard Randolph. *Under Barnegat's Beam: Light on Happenings along the Jersey Shore*. Privately Printed. Distributed by Appleton, Parsons & Co., New York, 1960, 43–46.

Kume, S., and J. Joseph. "Size Composition and Sexual Maturity of Billfishes Caught by the Japanese Longline Fishery in the Pacific Ocean East of 130 degrees." *Bulletin of the Far Seas Research Laboratory* 2 (1969): 115–162.

Lam, V. W. *et al.* "Projected change in global fisheries revenues under climate change." *Sci. Rep.* 6, 32607; doi: 10.1038/srep 32607 (2016).

Lloyd, John Bailey. *Six Miles at Sea—A Pictorial History of Long Beach Island, NJ*. Long Beach Island, NJ: Down the Shore Publishing and The SandPaper, 1990, 97–98.

Mackenzie, C. "Swordfishing from Martha's Vineyard." *Dukes County Intelligencer* 35, no. 3 (1994). Dukes County Historical Society, 101, 111, 115, 130.

"Marine Debris Biodegradation Time Line." *Mote Marine Laboratory Fact Sheet*, 1993. At 1600 Ken Thompson Parkway, Sarasota, FLA 37236.

McKeever, W. *Emperors of the Deep*. New York: Harper Collins, 2019, 40-65.

National Coalition for Marine Conservation Bulletin 2000, Leesburg, VA, 3, 5, 6, 8.

"National Report of the United States: 2000." *ICCAT Working Document SCRS/00/142* (rev.), US Department of Commerce, National Oceanic and Atmospheric Administration, National Marine Fisheries Service, October 2000, 2, 8, 11, 44.

O'Hara, K., S. Idicello and J. Zillisen. *A Citizens Guide to Plastics in the Ocean: More than a Little Problem*. Washington DC: Center for Marine Conservation, 1994, 15–16.

Our Living Oceans: Report on the Status of US Living Marine Resources, 1999. US Department of Commerce, National Marine Fisheries Service. NOAA Tech Memo NMFS-F/SPO-41, 301, 3, 101, 119, 121,188.

Palmer, J. Mark. "New US 'Dolphin-Safe' Label Is A Dolphin Death Certificate." *Earth Island Journal* 15, no. 2 (Summer 2000): 8, 1p, 1c.

Pierce, Wesley George. *Goin' Fishin': The Story of the Deep-Sea Fishermen of New England*. Salem, MA: Marine Research Society, 1934, 213–15.

Rigney, Matt. *In Pursuit of Giants: One Man's Global Search for the Last of the Great Fish*. New York: Viking Press, 2012, 94 - 95.

Robins, C. Richard, G. Carleton Ray, and John Douglass. *Atlantic Coast Fishes of North America. Peterson's Field Guides.* New York: Houghton Mifflin, 1986, 154–55, 262, 266.

Safina, Carl. *Song for the Blue Ocean.* New York: Henry Holt, 1997, 8, 13, 17, 20.

Shuda, Bill (captain). 1992. Homeport Charts, PO Box #730, Rio Grande, NJ.

Skerrett, P. J. "Designing Better Tankers." *Technology Review* 95, no. 6 (1997) Aug./Sept. 92: 8, 3p, 2c.

Smith, Jerome V. C. *Natural History of the Fishes of Massachusetts: Embracing a Practical Essay on Angling.* Boston: Allen and Ticknor, 1833, 311, 313, 315.

Smith, Russell. "The Sinking of the Andrea Doria." *Sweetwater Reporter News,* 1990. Online: http://camalott.com/~rssmith/Doria.htm.

Stock Assessment and Fishery Evaluation (SAFE) Report 2018 for Atlantic Highly Migratory Species. Chapter 6: Economic Status of HMS Fisheries, Table 6.3, Estimate of the total ex-vessel revenues of Atlantic HMS fisheries in 2010 – 2017, Commercial Fisheries, page 144. Source: National Oceanic and Atmospheric Administration, National Marine Fisheries Service, 2018. Online: http://www.nmfs.noaa.gov.

Stock Assessment and Fishery Evaluation (SAFE) Report 2018 for Atlantic Highly Migratory Species. Chapter 5: Fishery Data, Table 5.5, Reported numbers of catch in the U.S. pelagic longline Fishery 2013 – 2017, page 71. Source: National Oceanic and Atmospheric Administration, National Marine Fisheries Service, 2018. Online: http://www.nmfs.noaa.gov.

Stock Assessment and Fishery Evaluation (SAFE) Report 2018 for Atlantic Highly Migratory Species. Chapter 5: Fishery Data, Table 5.6, Reported landings (mt ww) in the U.S. Atlantic pelagic longline fishery in 2013- 2017, page 71. Source: National Oceanic and Atmospheric Administration, National Marine Fisheries Service, 2018. Online: http://www.nmfs.noaa.gov.

Stock Assessment and Fishery Evaluation (SAFE) Report 2018 for Atlantic Highly Migratory Species. Chapter 5: Fishery Data, Table 5.58, U.S. catches and landings (mt ww) of Atlantic swordfish by area and gear in 2013 – 2017 NW Atlantic, page 129. Source: National Oceanic and Atmospheric Administration, National Marine Fisheries Service, 2018.
Online: http://www.nmfs.noaa.gov.

Stock Assessment and Fishery Evaluation (SAFE) Report 2016 for Atlantic Highly Migratory Species. Section 5: Fishery Data-Pelagic Longline, Table 5.2, U.S. vs. Total International Catch of HMS Reported to ICCAT (Calendar Year 2015), page 56. Source: National Oceanic and Atmospheric Administration, National Marine Fisheries Service, 2016.
Online: http://www.nmfs.noaa.gov.

Stock Assessment and Fishery Evaluation (SAFE) Report 2001 for Atlantic Highly Migratory Species. Section 4: Fishery Data Update, Table 4.1, Calendar Year 1999 U.S. vs International Catch of HMS (mt ww) other than sharks. Source: NMFS, 2000a, U.S. Department of Commerce, National Oceanic and Atmospheric Administration, National Marine Fisheries Service, 2000, 4-2. Online: http://www.nmfs.noaa.gov.

Stock Assessment and Fishery Evaluation (SAFE) Report 2000 for Atlantic Highly Migratory Species. Section 4.1.3: U.S. vs. International Catch, Table 4.1.2, Estimated International Longline Landings in the Atlantic and Mediterranean: 1995-1998 (mt ww)*. Source: 1999 SCRS Report, US National Report, US Department of Commerce, National Oceanic and Atmospheric Administration, National Marine Fisheries Service, 2000. Online: http://www.nmfs.noaa.gov.

Ulanski, Stan. *The Billfish Story: Swordfish, Sailfish, Marlin, and other Gladiators of the Sea.* Georgia: The University of Georgia Press, 2013, 112 - 114,116.

Williams, M. *The Boater's Weather Guide.* Centerville, MD: Cornell Maritime Press, 1990, 67–69, 106, 134.

Young, Jock & Drake, Anita. (2004) Age and growth of broadbill swordfish (Xiphias *gladius*) from the Australian waters.
The Commonwealth Scientific and Industrial Research Organisation | CSIRO · Division of Marine and Atmospheric Research

Interviews

John Gallimore, former commercial fisherman. Telephone interview.

Saul Phillips, president, Export Inc. Personal interview.

Nelson Beideman, executive director, Blue Water Fishermen's Association. Personal interview.

Peter Dolan, longline captain, Barnegat Light, NJ. Personal interview.

J. Tim Hobbs, fisheries director, National Coalition of Marine Conservation. Telephone interview.

John Larson, Viking Village. Personal interview.

Keith Larson, former swordboat captain, Viking Village. Personal interview.

McCready, petty officer, U.S. Coast Guard Station, Barnegat Light, NJ. Telephone interview.

Holmes, chief warrant officer, U.S. Coast Guard Group, Atlantic City, NJ. Telephone interview.

Margaret Kearney, sales, Concorde Aerosales Inc. (immersion suites), Fort Lauderdale, FL. Telephone interview.

LCDR Frank Genco, USCG. Personal communication, e-mail.

Shawn Garrity, owner of the Lusty Lobster Seafood Market. Telephone interview.

Internet References

http://www.uscg.mil/d1/newengland/mfscmastatement.htm

"Statement of Rear Admiral George N. Naccara, Commander, First Coast Guard District on the Magnuson-Stevens Fishery Conservation and Management Act Before the Subcommittee on Oceans And Fisheries Committee on Commerce, Science, and Transportation, United States Senate, Boston, Massachusetts."

https://www/uscg/mil/datasheet/hh-65.htm

"HH-65A 'Dolphin' Short Range Recovery Helicopter."

https://www/uscg.mil/datasheet/hh-60.htm

"HH-60 'Jay Hawk' Medium Range Recovery Helicopter."

https://www.uscg.mil/datasheet/hc-130.htm

"HC-130 'Hercules' Long Range Surveillance Aircraft."

https://www.uscg.mil/hq/g-o/g-opr/amver/surpic.htm

"The Product of Amver. A Surface Picture (Surpic) of an Area of the Ocean."

https://www.uscg.mil.hq/g-cp/history/faqs/lss motto.html

"USCG FAQS from the Historian's Office. Origin of the saying: 'You have to go out, but you do not have to come back.'"

https://www.uscg.mil/datasheet/110wpb.htm

"USCG 110-Foot Patrol Boat (WPB) Island Class."

https://www.uscg.mil/datasheet/87wpb.htm

"USCG 87-Foot Coastal Patrol Boat (WAB) Marine Protector Class."

https://marinebio.org/conservation/sustainable-fisheries/
MarineBio Conservation Society

https://um.cfsan.fda.gov/ndms/mercury.html
"Mercury in Fish: Cause for Concern." US FDA for Consumer September 1994, revised May 1995.

https://fwie.fw.vt.edu/www/macsis/fish.htm
Marine Coastal Species Information Systems. Swordfish 26, Aug. 96 Bluefin Tuna, 1, 2, 4, 5.

https://abcnews.go.com/wire/World/reuters20010125_2768.html
"More than 70 Die Each Day Fishing at Sea—UN's FAO," reported on *World News Tonight*, January 25, 2001.

https://www.naia@naiaonline.org
National Animal Interest Alliance: "The tuna-dolphin connection."

https://www.fisheries.noaa.gov/national/marine-mammal-protection/dolphin-safe

https://www.mmc.gov/wp-content/uploads/2006annualreport.pdf
Gosliner, Michael L., 1999. The Tuna-Dolphin Controversy, 116 *Conservation and Management of Marine Mammals*, John R. Twiss, Jr. and Randall R. Reeves (eds.), Smithsonian Institution Press, Washington, D.C.

http://www.cnie.org/nle/mar-14.html
96011: Dolphin Protection and Tuna Seining
"CRS Issue Brief for Congress," by Eugene H. Buck, Environment and Natural Resources Policy Division: updated August 29, 1997.

http://unmuseum.mus.pa.us/squid.htm

Krysteh, L., ed. "The Giant Squid," 1996.

https://titanic.com/the-rescue.php3

"The Official Page of the Titanic: The Rescue."

https://www.pbs.org/lostliners/andrea.html

"Lost Liners and the Sinking of the Andrea Doria"

https://www.almanac.com/content/weather-sayings-and-their-meanings

"Weather Folklore."

https://www.fisheries.noaa.gov/species/north-atlantic-swordfish

North Atlantic Swordfish - Species Directory

www.garretthardinsociety.org/articles/art_tragedy_of_the_commons.html

The Tragedy of the Commons

Garrett Hardin, 1968

Published in Science, December 13, 1968.

https://www.researchgate.net/institution/The_Commonwealth_
Scientific_and_Industrial_Research_Organisation/CSIRO - Division of
Marine and Atmospheric Research

Abstract: Young, Jock & Drake, Anita. (2004) Age and growth of broadbill
swordfish (Xiphias gladius) from the Australian waters.

https://www.pewtrusts.org

The Pew Charitable Trusts

Research & Analysis

"Deep-Set Buoy Gear: A Better Way to Catch Swordfish," November 9, 2015.

https://www3.weforum.org/docs/WEF_Harnessing_4IR_Oceans.pdf

Insight Paper from *Harnessing the Fourth Industrial Revolution for Oceans*

World Economic Forum

November 2017.

https://population.un.org/wpp/DataQuery/

United Nations

Department of Economic and Social Affairs

Population Division

World Population Prospects 2017.

https://www.npg.org/library/forum-series/impact-us-pop-growth-climate-change.html

NPG Negative Population Growth Forum Paper

The Impact of U.S. Population Growth on Global Climate Change

Edwin S. Rubenstein

January 2017.

https://vikingvillage.homestead.com

Viking Village and Marina

Barnegat Light, NJ 08006.

Marine Conservation Organizations

Alliance for a Living Ocean (ALO)
https://www.livingocean.org

Animal Welfare Institute
https://awionline.org

Clean Ocean Action
https://www.cleanoceanaction.org

Earth Island Institute
https://www.earthisland.org

Greenpeace
https://www.greenpeace.org

Marine Mammal Stranding Center (MMSC)
https://www.mmsc.org

Natural Resource Defense Council (NRDC)
https://www.nrdc.org

OCEANA
https://www.oceana.org

Ocean Conservancy
https://www.oceanconservancy.org

SandyHook SeaLife Foundation (SSF)
https://www.sandyhooksealife.org

Sea Shepherd Conservation Society
https://www.seashepherd.org

SeaWeb
https://www.seaweb.org

Shark Research Institute (SRI)
https://www.sharks.org

Surfrider Foundation
https://www.surfrider.org

The 5 Gyres Institute
https://www.5gyres.org

The Environmental Defense Fund
https://www.edf.org

The International Commission for the Conservation of Atlantic Tuna (ICCAT)
https://www.iccat.int/en

The Pew Charitable Trusts

https://www.pewtrusts.org

U.S. Department of Commerce Office of Sustainable Fisheries

National Oceanic and Atmospheric Administration (NOAA)

National Marine Fisheries Service (NMFS)

https://www.nmfs.noaa.gov/Commercial.htm

Wild Oceans: for the future of fishing

Originally—National Coalition for Marine Conservation (NCMC)

https://wildoceans.org

On Fish and the Ocean

Bates, A. ©2020

Dark Side of the Ocean: The Destruction of Our Seas, Why it Matters, and

What We Can Do About It (Planet in Crisis)

GroundSwell Books

Block, B. ©2019

The Future of Bluefin Tunas: Ecology, Fisheries Management, and

Conservation

Johns Hopkins University Press

Bolster, W. Jeffrey ©2014

The Mortal Sea

First Harvard University Press paperback edition

Earle, S. and McKibben, B. ©2010

The World is Blue: How Our Fate and the Oceans are One

National Geographic

Earll, B. ©2018
Marine Conservation: People, Ideas and Action (Perspectives in
Conservation Biology)
Pelagic Publishing

Ellis, R. ©2013
Swordfish: A Biography of the Ocean Gladiator
University of Chicago Press

Ellis, R. ©2007
The Empty Ocean
Island Press

Greenlaw, L. ©2000
The Hungry Ocean: A Swordboat Captain's Journey
Hachette Books

Gronbaek, L./Lindroos, M./Munro, G./Pintassilgo, P. ©2020 (1st Edition)
Game Theory and Fisheries Management: Theory and Applications
Springer Press

Helfman, G. S. ©2007
Fish Conservation: A Guide to Understanding and Restoring Global Aquatic
Biodiversity and Fishery Resources
Island Press (1st Edition)

Henseler, L. and Henseler, C. ©2020

Advice from the Ocean: Unexpected Paths into Marine Conservation

Independently Published

Hilborn, U./ Hilborn, R. ©2019

Ocean Recovery: A Sustainable Future for Global Fisheries?

Oxford University Press

Holmes, G. ©2012

East of the Hague Line

Trafford, First Edition

Jackson, J. and Alexander, K. ©2011

Shifting Baselines: The Past and the Future of Ocean Fisheries

Island Press

Junger, S. ©2009

The Perfect Storm: A True Story of Men Against the Sea

W.W. Norton & Company

Longo, S./Clausen, R./Clark, B. ©2015

The Tragedy of the Commodity: Oceans, Fisheries, and Aquaculture

(Nature, Society, and Culture)

Rutgers University Press

Lynch, J. and Hall, W. ©2012

Sword: Harpooning Swordfish off the New England Coast, and its Demise

Gaelforce Press LLC

McKeever, W. ©2019

Emperors of the Deep: Sharks – The Ocean's Most Mysterious, Most Misunderstood and Most Important Guardians

Harper Collins Books

Pauly, D. ©2019

Vanishing Fish: Shifting Baselines and the Future of Global Fisheries

Greystone Books

Roberts, C. ©2013

The Ocean of Life: The Fate of Man and the Sea

Penguin Books

Roscam Abbing, M. ©2019

Plastic Soup: An Atlas of Ocean Pollution

Island Press

Safina, C. ©1999

Song for the Blue Ocean

Holt Paperbacks 1st Owl Books

NOAA Technical Report NMFS Circular 441

Synopsis of the Biology of the Swordfish, *Xiphias gladius* Linnaeus

B. J. Palko, G. L. Beardsley, and
W. J. Richards

November 1981

FAO Fisheries Synopsis No. 127

U.S. DEPARTMENT OF COMMERCE
Malcolm Baldrige, Secretary

National Oceanic and Atmospheric Administration
John V. Byrne, Administrator

National Marine Fisheries Service

249

CONTENTS

---

*No information available.

Synopsis of the Biology of the Swordfish, *Xiphias gladius* Linnaeus[1]

R. J. PALKO, G. L. BEARDSLEY, AND W. J. RICHARDS[2]

1 IDENTITY

1.1 Nomenclature

1.11 Valid name

Xiphias gladius Linnaeus, 1758 (Fig. 1).

Originally described by Linnaeus in 1758. The type locality was listed as "European ocean." No type specimen. (Linnaeus 1758. Systema Naturae Vol. 10, p. 248.)

1.12 Objective snyonymy

Xiphias gladius Linnaeus, 1758 (see above).

Xiphias imperator Bloch and Schneider, 1801. Type locality: Mediterranean Sea. Type specimen: none.
Xiphias rondeletii Leach, 1818. Type locality: Queensferry, Scotland. Type specimen: a stuffed specimen at the College at Edinburgh (present status unknown).
Phaetonichthys tuberculatus Nichols, 1923. Type locality: Rapa Island, Austral Group, South Pacific. Type specimen: American Museum of Natural History 8257, mutilated specimen, tail only.
Xiphias thermaicus Serbetis, 1951. Type locality: Gulf of Thermaicus, Mediterranean Sea. Type specimen: none.

Of these names in the objective synonymy, some comment is deserved. *Xiphias imperator* Bloch and Schneider is illustrated as having pelvic fins which *X. gladius* lacks.
Ziphias gladius Hector, 1875 has been listed in other synonymies (Nakamura et al. 1968), but our reading of that citation failed to reveal the use of that name. We expect it may be in some other publication, but simply represents a misspelling of *Xiphias. Phaetonichthys tuberculatus* Nichols was described from only the tail of a specimen found in the gullet of a tropic bird. It bears the features of a juvenile swordfish. The description of *X. thermaicus* also appears to be a young speciment 0.30 m in length.

1.2 Taxonomy

1.21 Affinities

Suprageneric

Phylum Chordata
Subphylum Vertebrata
Superclass Gnathostomata
Class Osteichthyes
Subclass Actinopterygii
Order Perciformes
Suborder Scombroidei
Family Xiphiidae

The above classification follows Greenwood et al. (1966), but Gosline (1968), in his review of perciform suborders, presented evidence that the relation of billfishes to scombrids and their allies may be one of convergence. He placed the monotypic family Xiphiidae along with the family Istiophoridae and, provisionally, the Luvaridae, in a separate suborder Xiphioidei.

Generic

Monotypic genus, see specific diagnosis.
Xiphias Linnaeus, 1758 (ref.). Type species: *Xiphias gladius* Linnaeus, 1758 by monotypy.

Figure 1.—Swordfish, *Xiphias gladius* Linnaeus, 1758.

[1]Southeast Fisheries Center Contribution No. 81-15m.
[2]Southeast Fisheries Center Miami Laboratory, National Marine Fisheries Service, NOAA, 75 Virginia Beach Drive, Miami, FL 33149.

Specific

Diagnosis: pelvic fins absent, scales absent in adult, one pair of caudal keels, snout long and sword shaped, somewhat flattened in cross section, base of first dorsal fin short and broadly separated from the second dorsal fin.

1.22 Taxonomic status

The most recent review is by Nakamura et al. (1968) and they consider the swordfish to be a cosmopolitan species of the monotypic family Xiphiidae.

1.23 Subspecies

None recognized.

1.24 Standard common names and vernacular names

The names capitalized are official or in more common use. Compiled from Rosa (1950) and Nakamura et al. (1968).

Algeria - pesce espada, pez espada
Argentina - PEZ ESPADA
Belgium - ESPADON
Brazil - PEIXE ESPADA, agulhão, espadarte
Ceylon - kadu koppara
Chile - PEZ ESPADA, Albacora
China - chien yu
Cuba - PEZ-ESPADA, emperador
Denmark - SVAERDFISK
France - ESPADON, emperadour, pei empereur
French West Africa - bongjhojh
Germany - SCHWERTFISCH
Greece - XIPHIAS
Ireland - luinniasc
Italy - PESCE SPADA
Japan - MEKAJIKI
Madeira Is. - PEIXE AGULHA
Malta - PISCISPAT, pixxi spad
Mexico - PEZ ESPADA
Netherlands - ZWAARDVISCH
New Zealand - BROADBILL
Norway - SVAERDFISK
Philippines - SWORDFISH; dugso, malasagi (Bikol dialect); malasugi, manumbuk - (marinao, samal and tao sug); dogso, lumod, malasugi, mayaspas - (visayan)
Portugal - AGULHÃO, agulha, espadarte, peixe agulha
Rumania - PESTE CU SPADA
Spain - PEZ ESPADA, espada, chichi spada, emperador, peix espasa, jifia, espasa, espardarte, aja pala
Sweden - SWARDFISK
Taiwan - ting mien chiu chi yu, pai jou ting pan
Tunisia - pesce espada, boussif
USSR - mesh-riba, meshvenosouiye
Union of South Africa - SWORDFISH, broadbill
United States - swordfish, broadbill, common swordfish, spearfish
Venezuela - PEZ ESPADA, emperador, espadon
Vietnam - ho ca mui kiem
Yugoslavia - SABLJAN, sabljack, iglun, jaglun, macokljun, pese spada.

1.3 Morphology

[Condensed from Nakamura et al. (1968) and Richards (1974).]

1.31 External morphology

Adult fish have no scales or teeth, but the young stages have pronounced atypical scales. There is no pelvic fin. The base of the first dorsal is short in the adults but very long in the young and confluent with the second dorsal; in the adult the two fins are widely separated. The snout is long in all sizes, broadly flattened in the adults. The body is rather heavy and round in the adults, but in the young it is long, thin, and snakelike. The cranium is hard and wide; temporal crest and pterotic crest in the posterior end of the cranium are fairly well developed. The posterior projection of the pterotic and the epiotic are weakly developed. The supraoccipital projection is fairly well developed. The neural and haemal spines of the vertebrae are flattened. There are 26, rarely 27 vertebrae, 15–16 trunk plus 10–11 caudal. [Richards (1974) reversed the vertebral counts.] The lateral apophysis is not well developed.

There are two dorsal fins with 38–49 rays in the first and 4–5 rays in the second. There are two anal fins: 12–16 rays in the first, 3–4 rays in the second. There are 17–19 pectoral rays and no pelvic fin.

A striking change in shape takes place during growth. The body not only changes from long and thin, but the fins, coloration, squamation, and bill shape also change radically. There are many recent accounts of swordfish ontogeny as well as accounts early in this century. Richards (1974) gave details of the history of swordfish development and cited the various accounts.

1.32 Cytomorphology

Nothing found in the literature.

1.33 Protein specificity

Nothing found in the literature.

2 DISTRIBUTION

2.1 Total area

Swordfish are the most widely distributed billfish and occur worldwide from about lat. 45°N to 45°S in all tropical, subtropical, and temperate seas (Fig. 2).

In the western Atlantic, swordfish occur from the coast of Newfoundland (Tibbo et al. 1961) to Argentina (de Sylva 1962; Wise and Davis 1973).

In the eastern Atlantic, swordfish have been recorded off Scandinavia (Duncker 1936), Great Britain (Rich 1947), France, Spain, in the Mediterranean Sea (Sanzo 1922), and the Sea of Marmara (Artüz 1963). Swordfish have been reported in the Black Sea off the coast of Bulgaria and Rumania, and there are indications that swordfish migrate from the Black Sea to the Sea of Azov in the summer (Ovchinnikov 1970). In the Baltic Sea, swordfish have been collected off Tallinn and Haapsalu (Ovchinnikov 1970) and Wolin Island (Jakuczun 1971). In the eastern South Atlantic, swordfish occur along the west coast of Africa down to the Cape of Good Hope (Penrith and Cram 1974).

In the eastern Pacific, swordfish range from Oregon (Fitch 1960) to Talcahuano, Chile (Lobell 1947). Swordfish are caught

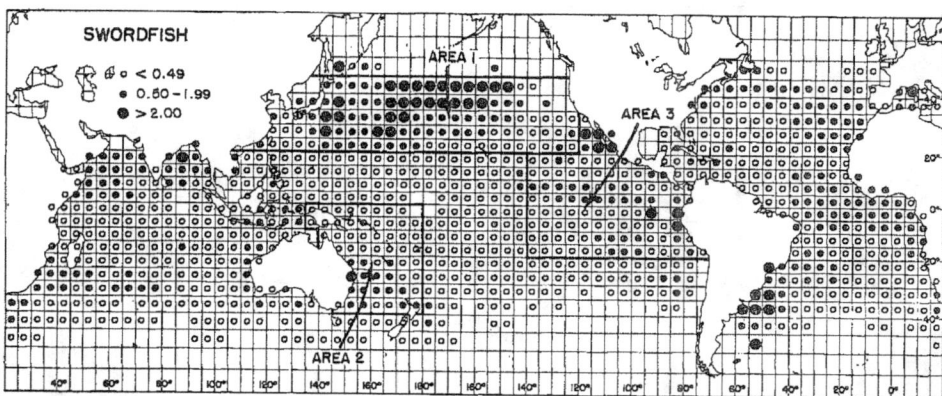

Figure 2.—Distribution of swordfish in the Pacific, Indian, and Atlantic Oceans based on catch rates from the Japanese tuna longline fishery. The circles indicate mean catch rates (number of fish per 1,000 hooks). Also shown (areas 1, 2, and 3) are centers of concentration of hypothesized swordfish stocks in the Pacific (from Sakagawa and Bell see text footnote 10, fig. 1.)

off the Hawaiian (Strasburg 1970) and Galapagos Islands. In the western Pacific, based on Japanese longline data, swordfish are widely distributed from temperate waters off the coast of Japan (Yabe et al. 1959) to the waters of Australia and New Zealand (Webb 1972).

Swordfish occur in the Indian Ocean with areas of concentration off the coasts of India (Chacko et al. 1964), Ceylon (Deraniyagala 1951), Saudi Arabia (Rass 1965), the east coast of Africa, and around the Cape of Good Hope. Nishikawa and Ueyanagi (1974) have also shown areas of concentration between lat. 20° and 40°S, long. 60° and 100°E. Adult swordfish also occur in good concentrations off the southwest coast of Australia.

2.2 Differential distribution

2.21 Spawn, larvae, and juveniles

Swordfish larvae occur in all the tropical seas, and their distribution is closely associated with surface temperatures between 24° and 29°C (Täning 1955). Gorbunova (1969) found concentrations of swordfish larvae in the southwestern area of the Atlantic, in the eastern part of the Indian Ocean, and in the central waters of the Pacific Ocean south of the equator.

In the western Atlantic, Markle (1974) stated, based on her own sampling plus similar data from other sources, that the greatest densities of swordfish larvae occur from the Straits of Florida to Cape Hatteras and in the Virgin Islands—Leeward Islands area (Fig. 3). Arata (1954) and Arnold (1955) suggested that the Gulf of Mexico serves as a nursery ground for swordfish.

In the eastern Atlantic, Gorbunova (1969) found swordfish larvae off the northwestern shores of Africa in the areas of subtropical divergence. Swordfish larvae also occur in the Mediterranean Sea (Sella 1911; Sanzo 1922) and in the Straits of Messina from June to September. Artüz (1963) presented evidence that swordfish in the Sea of Marmara spawn in coastal waters because eggs and juveniles are found there in April, May,

and June. He believed this was a separate stock from those found in the Mediterranean.

Swordfish larvae are widely distributed in the tropical central and western Pacific (Fig. 4), and this distribution pattern appears to be governed by the position of the 24°C surface isotherm (Nishikawa and Ueyanagi 1974).

In the Indian Ocean, swordfish larvae are distributed throughout the area southwest of Sumatra (Bogorov and Rass 1961) and are abundant between lat. 12° and 17°S, approximately in the area of the South Equatorial Current and in the eastern section along long. 85° and 87°E between lat. 5°N and 17°S.

See also section 3.16.

2.22 Adults

In the Atlantic Ocean, relatively little information is available on the seasonal distribution of swordfish. The United States and Canadian longline and harpoon fisheries operate in the western North Atlantic in summer and early fall. In winter, most longline activity moves to more southern waters. Guitart-Manday (1975) noted that swordfish were caught throughout the year off Cuba, but best catches were in the winter and early spring and consisted of mostly large females while catches in summer were few and consisted mostly of males. Recently, however, substantial concentrations of swordfish have been discovered in the Gulf of Mexico and off the east coast of Florida in the summer months. Tagging data shed little light on seasonal distribution. Results of Canadian tagging experiments have shown that all recaptures of tagged swordfish have been made within the same general area of release and at the same season of the year (Beckett 1974). One swordfish tagged in the Gulf of Mexico in March 1974, however, was recaptured off Georges Bank in August 1977, indicating that at least some swordfish make substantial seasonal migrations (Casey 1977[3]).

Joseph et al. (1974) reported areas of good fishing for sword-

---

[3]Casey, J. 1977. The shark tagger, winter 1977. Newsletter of the Cooperative Shark Tagging Program, 8 p. U.S. Dep. Comm., NOAA, NMFS, Northeast Fisheries Center, RR7A, Box 522A, Narragansett, RI 02882.

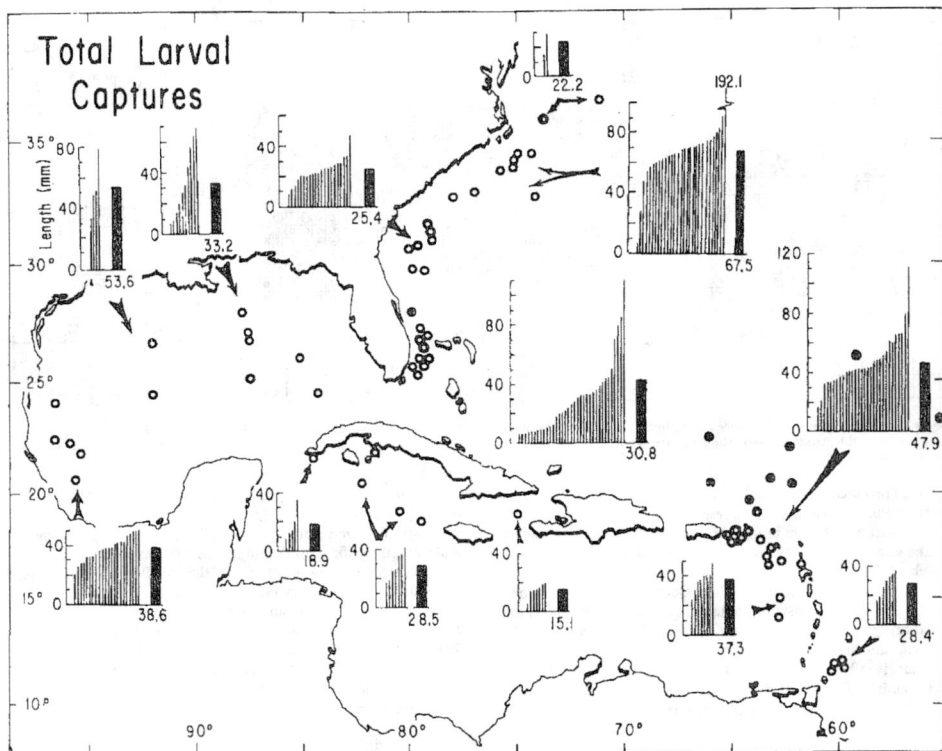

Figure 3.—Numbers, size ranges, and mean lengths of swordfish larvae—collections from various sources including Fisheries Research Board of Canada (from Markle 1974, fig. 3 and caption).

fish between lat. 35°N and 40°S, with best catches in the coastal areas between lat. 20° and 30°N and the areas adjacent to Baja California. For the southeast Pacific, the principal swordfish grounds are centered in coastal waters from the equator to about lat. 15°S and around the Galápagos islands. Concentrations of swordfish extend westward from this area in a longitudinal band along the equator during all seasons. During the first and fourth quarters (southern spring and summer), a secondary longitudinal band extends westward between lat. 10° and 20°S. Royce (1957) believed that the distribution of swordfish in the western Pacific indicates that the adults prefer cooler waters than other billfishes.

In the Indian Ocean, concentrations of swordfish occur off Saudi Arabia at lat. 15°N, and long. 40°-70°E and in the Bay of Bengal between long. 90° and 95°E (Rass 1965; Nakamura 1974).

2.3 Determinants of distribution changes

Their geographical distribution in the northwestern Atlantic apparently varies considerably due to marked seasonal variation in environmental conditions (Beckett 1974). In the winter, swordfish are confined to waters associated with the Gulf Stream, but in the summer they are found over a much wider area. Feeding habits and temperature variations apparently influence a differential distribution by size with larger fish being found in cooler water and few fish under 90 kg seen in waters of <18°C. Sex ratios also differ with temperature, as few males are found in the colder (below 18°C) water.

There is undoubtedly a relationship between the occurrence of swordfish larvae and the distribution, interaction, and modification of water masses according to Tibbo and Lauzier (1969). All larvae they examined were captured within relatively narrow ranges of surface temperature (23.4°-25.6°C) and salinity (35.81-36.36°/oo).

In the Pacific, distribution of larval swordfish is associated with the North Equatorial Current or the Kuroshio Current during April and May (Nakamura et al. 1951). Nishikawa and Ueyanagi (1974) noted a marked difference between surface and subsurface catches during the day but not as much during the night and felt this difference represented diurnal vertical movements of larval swordfish. In the Pacific, swordfish larvae and juveniles are restricted to areas of upwelling where high productivity provides favorable conditions (Gorbunova 1969). Yabe et al. (1959) stated that young swordfish are distributed in the tropical and subtropical zones and migrate to

256

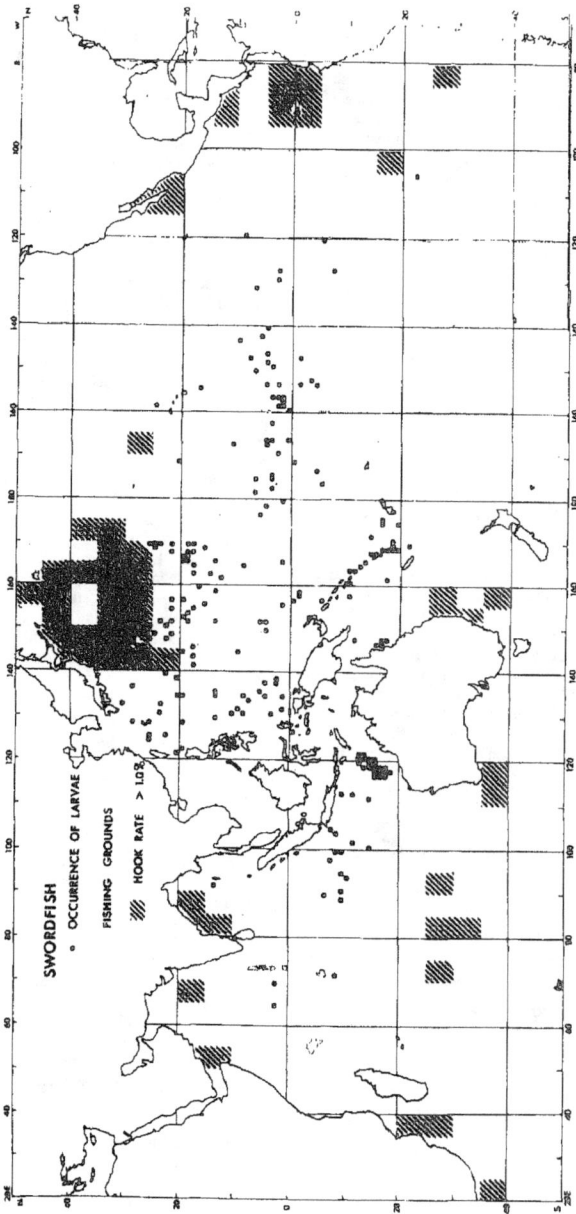

Figure 4.—Distribution of swordfish larvae (dots) and adults (hatched) in the Pacific and Indian Oceans. The adult distribution is represented by areas in which longline catches averaged greater than 1.0 fish per 1,000 hooks during 1970, and where fishing effort exceeded 20,000 hooks fished (from Nishikawa and Ueyanagi 1974, fig. 3 and caption.)

257

higher latitudes as their size increases, and Gorbunova (1969) indicated that juvenile swordfish do not migrate far during their first year of growth.

Kondritskaya (1970) noted that the distribution of swordfish larvae in the Indian Ocean is bounded by the 24 °C mean annual isotherm. In the northwestern Indian Ocean, Osipov (1968) believed that catches of swordfish change from area to area and show a seasonal variation between summer and winter, which is related to environmental conditions or eddy patterns within major current systems such as the Monsoon Current and the Equatorial Counter-current.

2.4 Hybridization

Nothing found in the literature.

3 BIONOMICS AND LIFE HISTORY

3.1 Reproduction

3.11 Sexuality

Swordfish are heterosexual. No known external characters distinguish males from females, although large swordfish are usually female.

3.12 Maturity

There is little information on size and age at first maturity and some of it is contradictory. Yabe et al. (1959) stated that swordfish first spawned at 5-6 yr and 150-170 cm eye-fork length. Ovchinnikov (1970) said males reached sexual maturity at a length of around 100 cm and females at a length of 70 cm (author's note: measurement parameters not given in translation); however, recent research conducted on swordfish off the southeast coast of the United States indicates that males mature at a smaller size than females (about 21 kg for males and 74 kg for females) (E. Houde[a]). Kume and Joseph (1969) treated swordfish of < 130 cm in eye-fork length as immature.

3.13 Mating

Swordfish are apparently solitary animals and rarely gather in schools. However, several occurrences of pairing thought to be associated with spawning have been noted in the Atlantic (Guitart-Manday 1964) and in the Mediterranean (LaMonte and Marcy 1941), although pairing is considered a rarity.

See section 3.16.

3.14 Fertilization

Fertilization is external.

3.15 Gonads

The testes are paired, elongate organs, thin and ribbonlike in immature fish and flattened, pinkish white, and not round in cross section in adults. The ovaries are paired, elongate, and always round in cross section even in immature fish. The ovaries are always

much shorter and thicker than the testes (Artüz 1963). Sella (1911) reported that the swordfish ovary contracts after spawning and remains compact and firm.

3.16 Spawning

Spawning generally takes place in tropical waters where surface temperatures are > 20°-22 °C. In the Atlantic, spawning apparently occurs throughout the year in the Caribbean, Gulf of Mexico, and in the waters off Florida with the peak of the spawning season from April through September (Arata 1954). Swordfish with ripe ovaries have been reported off Cuba during the winter months (LaMonte 1944; Guitart-Manday 1975). In the same months, in the waters near Cuba, the Gulf of Mexico, the Caribbean Sea, and around the Cayman Islands, Guitart-Manday (1975) found many swordfish larvae and juveniles.

Tibbo and Lauzier (1969) concluded from their samples and the direction and speed of surface currents that larvae caught in the Florida Straits and off Cape Hatteras were hatched in the southern Gulf of Mexico and probably in the Yucatan Channel. They also believed that larvae taken near the Virgin Islands most probably originated in the southernmost part of the Sargasso Sea and possibly at the northern edge of the Equatorial Current. The sizes also indicated spawning from mid-December to mid-February.

Additional spawning areas have been reported—one in the Mediterranean Sea off Sicily and the Straits of Messina from June through August with peak spawning in the early July (Sella 1911; Sanzo 1922; Cavaliere 1963) and the other in the Sea of Marmara where there is some evidence that swordfish spawn in coastal waters in April, May, and June (Artüz 1963).

In the eastern Pacific, fish about to spawn are found in every month of the year but appear to be most abundant from March through July in northern latitudes and around January in southern latitudes (Kume and Joseph 1969). Although Yabe et al. (1959) assumed that all ripe ova were spawned at one time, Uchiyama and Shomura (1974) indicated that partial spawning could not be discounted since they found one ripe ovary which contained residual ova from an earlier spawning.

Matsumoto and Kazama (1974) believed that there was evidence of a difference in spawning time in various parts of the Pacific, which was reflected by the seasonality of occurrence of larvae and juveniles. Spawning occurs in spring and summer (March-July) in the central Pacific and in spring (September–December) in the western South Pacific. They noted that spawning takes place all year in equatorial waters and begins and ends 1 or 2 months earlier in the western Pacific in the Philippine-Formosa areas compared with the Hawaiian Islands area. Tsi-Gen (1960) stated that spawning individuals could be found at any time of the year in the western Pacific. Yabe et al. (1959) found a spawning area south of the subtropical convergence which they believed formed the recruitment for the fishing grounds of the North Pacific. They also collected mature ovaries in the northern part of the Coral Sea in October and in the Fiji Islands in June.

In the Indian Ocean, Yabe et al. (1959) found ripe ovaries in April and young swordfish in the lower latitude areas near the equator in August, November, and December. Based on swordfish larvae found in stomach contents of tunas and marlins, they concluded that the spawning season started after April and continued until December. In the Mozambique Channel, a swordfish larva was collected in late January (Kondritskaya 1970) and larvae have been reported east of Madagascar Island (Lütken 1880; Gorbunova 1969), although no dates are available.

[a]Dr. Ed Houde, Rosenstiel School of Marine and Atmospheric Science, 4600 Rickenbacker Causeway, Miami, FL 33149, pers. commun. June 1979.

3.17 Spawn

Uchiyama and Shomura (1974) classified the ova into several developmental states that are not dependent on ova diameters:

1. Primordial ova—Ova transparent, ovoid, and diameters range from 0.01 to 0.05 mm. Primordial ova are present in all ovaries.
2. Early developing ova—Ova still transparent and ovoid; diameters range from approximately 0.06 to 0.24 mm. A chorion membrane has developed around the ovum and opaque yolklike material has begun to be deposited within the ovum.
3. Developing ova—Ova completely opaque, more wedge-shaped than ovoid, and diameters range between 0.16 and 0.96 mm. The chorion is stretched and not visible in this stage.
4. Advanced developing ova—Ova ovoid and diameters range from 0.47 to 1.20 mm. Ova have a translucent margin, a fertilization membrane, and a round yolk.
5. Early ripe ova—Ova diameters range from 0.60 to 1.20 mm. The yolk material is translucent and oil globules have begun to form.
6. Ripe ova—Ova transparent and with oil globules. Diameters range from 0.80 to 1.66 mm.
7. Residual ova—Ova in this stage show signs of degeneration and ova are thin-walled, translucent, and measure approximately 0.80 mm in diameter.

According to Sella (1911) and Sanzo (1922), "the eggs are buoyant and transparent with a large oil droplet. The yolk sac is very small; yolk vesicular; capsule with a quite evident amorphous network. The egg diameter is about 1.6–1.8 mm; the oil droplet, 0.4 mm."

Fish (1926) noted that a swordfish weighing 68 kg contained maturing ovaries weighing 1.5 kg and estimated the number of ova at 16,130,400, measuring ova only from 0.1 mm to 0.55 mm in diameter.

3.2 Preadult phase

3.21 Embryonic phase

Embryonic development is rapid (2½ days of incubation or a little more). At an early stage, melanophores are over all the surface of the yolk and trunk. After 1 day of incubation, they are already well dispersed, and at this stage there begins to appear on the trunk a diffused straw-yellow color. The development of the chromatophores on the yolk and their branching render the egg less transparent and a dirty-white color. At the beginning of the third day, the melanophores are again augmented, and those on the trunk, corresponding to the segments, have stronger ramification. The diffused straw color is intensified. There are 24 segments (Padoa 1956).

3.22 Larval phase

See section 3.23

3.23 Adolescent phase

Like other billfishes, swordfish gradually metamorphose from a larva to an adult (Fig. 5); therefore, all descriptive phases will be discussed as one continuous transformation.

Sanzo (1909, 1910, 1922, 1930) described eggs, described eggs and larvae at hatching, reared larvae from eggs through the yolk sac stage, and described a 13 mm and a 6 mm specimen. Sella (1911) confirmed Sanzo's (1910) work. Yasuda et al. (1978) described in detail embryonic and larval development of swordfish from the Mediterranean. Their results differed slightly from Sanzo (1910) and Sella (1911), particularly in the numbers of myomeres and in total length in the most advanced larval stages.

Cuvier (in Cuvier and Valenciennes 1831) was the first to describe a young swordfish and include a figure of a juvenile. In 1880 both Günther and Lütken (separate papers) described young specimens. Goode (1883) reviewed all previous work and added another descriptive note by Steindachner (1868) who described two young subadults. Regan (1909) pointed out the resemblance of a young *Xiphias* (200 mm in length) to the fossil species *Blochius longirostris*. Regan (1924) described and figured this 200 mm juvenile and noted that *Phaethonichthys tuberculatus* Nichols (1923) is actually a young swordfish and placed in the synonymy of *X. gladius*. Fowler (1928) also figured a young swordfish (ca. 225 mm) and like Regan (1924) noted that *P. tuberculatus* Nichols was a synonym of *X. gladius*. Therefore, by early in the century the young stages of swordfish were well described. Later accounts, which include descriptions of young swordfish, are Nakamura et al. (1951), Yabe (1951), Arata (1954), Tåning (1955), Jones (1958), Yabe et al. (1959), Gorbunova (1969), and Tibbo and Lauzier (1969[1]).

According to Richards (1974) there is no problem in separating young swordfish from istiophorids since swordfish lack the strong pterotic and preopercular spines which are so prominant in the early stages of other billfishes. In sizes over 20 mm, young swordfish are very dissimilar to other billfishes in appearance.

Swordfish larvae are easily recognized by their long snouts, heavily pigmented elongate bodies, and prominent supraorbital crest. Above 8.0 mm they have one or more rows of spinous scales on each side of the dorsal and anal fins, with those along the latter continuing forward to the level of the pectoral fin (Matsumoto and Kazama 1974).

Arata's (1954) observations on color of a live 68.8 mm swordfish are as follows: "The over-all color of the dorsal surface . . . was royal blue marked alternately with seven vertical bands of light blue from the head to the caudal fin . . . . The bands were not uniform in width and did not become silver until the ventral-most one-fifth of the lateral aspect was reached. The first band was located behind the head and the last band reached from midway on the caudal peduncle onto the caudal fin. The lower edge of the anal was hyaline, as was the posterior edge of the dorsal. Both tips of the caudal were also hyaline and the pectoral showed no color. The distal one-fourth of the premaxillary was white, and the extreme tip of the mandible lacked color. In profile only the extreme edge of the ventral aspect was silver, continuing from the tip of the mandible all the way to the caudal base. The dorsal fin was dark blue (almost black) with light blue areas corresponding to the light bars on the body. The anal fin lacked any carryover of pigment. There was a dark blue bar running obliquely from the gape (ahead of the angle of the mouth)

[1]Tibbo, S. N., and L. M. Lauzier. 1969. On the origin and distribution of larval swordfish *Xiphias gladius* L. in the Western Atlantic. Fish. Res. Board Can. Tech. Rep. 136, 20 p. Department of Environment, Fisheries and Marine Service, Office of the Editor, 116 Lisgar St., Ottawa, Canada K1A 0H3.

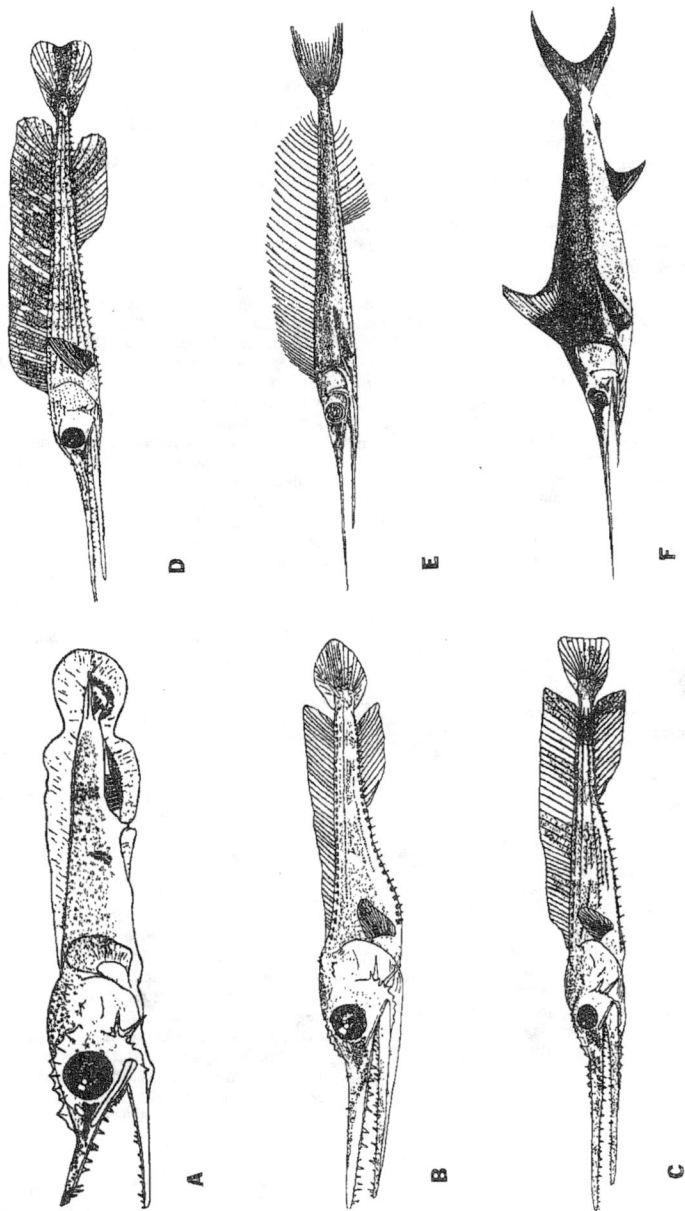

Figure 5.—Larvae and young of the swordfish: (A) 7.8 mm SL; (B) 14.5 mm SL; (C) 27.2 mm SL; (D) 68.8 mm SL; (E) 252 mm BL (length from posterior edge of orbit to base of caudal fin); (F) 580 mm BL. Specimens A, B, C, and D from Arata (1954); E and F from Nakamura et al. (1951).

260

through the middle of the eye and was lost about three-fourths of the way across the opercle."[1]

3.3 Adult phase

3.31 Longevity

Based on tagging data from 1961 to 1976, the maximum age is at least 9 yr, assuming it takes a minimum of 2 yr for a swordfish to grow large enough to tag (Beardsley 1978[6]).

There are insufficient data to determine whether the greater size attained by females relative to males is due to a more rapid growth rate or to a considerably longer life span (Beckett 1974). No information is available for the Pacific and Indian Oceans, although Beardsley (see footnote 6) noted that age-weight data suggest Pacific stocks may have a slower growth rate and grow to a smaller asymptotic size than Atlantic swordfish.

3.32 Hardiness

Adult swordfish must be adaptable to relatively large changes in their environment. They make feeding excursions into waters of 5°–10°C and depths of at least 650 m (Church 1968). They are able to make horizontal migrations from tropical and subtropical zones to the temperate waters in all oceans of the world. They have been sighted at the head of submarine canyons in the Gulf of Mexico in depths in excess of 455 m (Church 1968). In addition, swordfish are adaptable in their ability to utilize a variety of food sources, feeding on both surface and near surface animals as well as benthic species and those species which occur in between.

3.33 Competitors

Swordfish presumably compete with other billfishes as well as other large pelagic predators for the same organisms in the food chain. There is no information, however, on the effect this competition has on their survival. That effect is probably minimal since swordfish are capable of feeding on a variety of foods from the surface to the floor of the oceans, can travel from tropical to temperate waters, and are opportunistic predators.

3.34 Predators

The larvae of swordfish are a common food source of other fishes, including larger swordfish. The juveniles and young swordfish presumably are also preyed upon by any sufficiently larger predacious fish. Yabe et al. (1959) found young swordfish in the stomachs of the following 10 predators: blue marlin, *Makaira nigricans*; black marlin, *M. indica*; sailfish, *Istiophorus platypterus*; yellowfin tuna, *Thunnus albacares*; albacore, *T. alalunga*; bigeye tuna, *T. obesus*; dolphin, *Coryphaena hippurus*; striped marlin, *Tetrapturus audax*; shortbill spearfish, *T. angustirostris*; and blue shark, *Prionace glauca*.

Adult swordfish have few known natural enemies. Sperm whales, *Physeter catodon*; killer whales, *Orcinus orca*; and large sharks are perhaps the only species capable of preying on adult

swordfish (Tibbo et al. 1961). Sharks are the only creatures ever seen in actual combat with swordfish (Stark 1960; Maksimov 1968; Bozhkov 1975). Bigelow and Schroeder (1953) reported finding a good sized swordfish in the stomach of a mako shark, *Isurus oxyrinchus*. Sharks are regularly known to attack swordfish that have been harpooned, hooked on longlines, or caught by sport fishermen.

3.35 Parasites, diseases, injuries, and abnormalities

Silas (1967) and Silas and Ummerkutty (1967) summarized records of parasites in swordfish (Table 1). Iles (1971) noted that the presence of distinct species of *Tristoma* on swordfish from Hawaii and the possible difference in numerical incidence of *T. coccineum* and *T. integrum* on swordfish from the Mediterranean suggest that these monogenetic trematodes may be useful as biological tags to distinguish populations of swordfish.

Swordfish in the northwest Atlantic are usually heavily parasitized with many species, including sea lampreys (Tibbo et al. 1961). Guitart-Manday (1964) found the following parasites in swordfish off Cuba: *Ascaris incurva* in almost all stomachs; unidentified cestoda attached to outer walls of the stomach; unidentified *Hirudinea* also found in the stomach; and an ectoparasite of the genus *Pennella* deeply inserted in the subcutaneous muscular tissue. Swordfish in the Sea of Marmara frequently have a parasitic copepod (family Pennellidae) attached. Numerous nematodes are found in their stomachs and two kinds of cestodes are found in the intestines (Artüz 1963).

In the eastern Pacific off Hawaii, Yamaguti (1968a) reported two monogenetic trematodes, *Tristoma adintegrum* and *T. adcoccineum*, as occurring in the gills of swordfish. Two digenetic trematodes, *Maccallumtrema xiphiados* and *Reniforma mutilobularis*, occur in the abdominal muscle and gill filaments (Yamaguti 1970). In addition, he noted two cestodes, *Pseudeubothrium xiphiados* and *Bothriocephalus manubriformis*, as occurring in the intestines (Yamaguti 1968b). Ho (1963) reported a parasitic copepod, *Gloiopotes longicaudatus*, occurring on the general body surface and ventral surface of the "sword" in the waters off Formosa. Many swordfish have marks of sea lampreys on them but few have open wounds (Tibbo et al. 1961). The marks consist mostly of longitudinal scratches along the side of the body and indicate that the swordfish is fairly successful in ridding itself of these pests. Guitart-Manday (1964) found one or more very shallow oval wounds which apparently did not affect the subadjacent muscular layers. He assumed that these wounds were caused by a species of Ciclostomata (sic Cyclostomata). More recent information (Jones 1971) has shown that these wounds were probably caused by a small species of squaloid shark, *Isistius brasiliensis*.

There is no information on disease associated with swordfish.

3.4 Nutrition and growth

3.41 Feeding

Gorbunova (1969) had one of the best descriptions of feeding behavior in larval swordfish. These larvae apparently take food items lying either slightly above or on the same level as themselves, but mainly the larvae of planktophages which also feed during daylight hours. The intensity of capture of food is increased in the morning and evening and reduced at midday and at night. According to Leshcheva (1967) (cited in Gor-

[6]Beardsley, G. L. (editor). 1978. Report of the Swordfish Workshop held at the Miami Laboratory, Southeast Fisheries Center, Miami, Fla. June 7-9, 1977. Coll. Vol. Sci. Pap., VII(SCRS-1977):149-158. Int. Comm. Conserv. Atl. Tunas, General Mola 17, Madrid 1, Spain.

Table 1.—List of parasites found on swordfish (adapted from Silas 1964 and Silas and Ummerkutty 1964).

Locality	Parasite	Location on host
Monogenetic trematodes (Silas 1967)		
Atlantic	*Capsala laevis*	
Adriatic, Mediterranean, N.W.		
Atlantic, Pacific	*Tristoma integrum*	
Atlantic	*Tristoma coccineum*	gills
Digenetic trematodes (Silas 1967)		
Atlantic—Woods Hole	*Didymocystis xiphoides*	muscle and gill cavity
Pacific, Atlantic	*Hirudinella clavata*	stomach
Atlantic	*Hirudinella ventricosa*	
Cestodes (Silas 1967)		
N.W. Atlantic, Europe	*Fistulicola plicatus*	walls of intestine and rectum
N.W. Atlantic	*Grillotia erinaceus*	stomach and intestine walls
N.E. Atlantic and Mediterranean	*Gymnorhynchus (Gymnorhynchus) gigas*	muscles
Atlantic (Woods Hole)	*Gymnorhynchus (Molicola) uncinatus*	muscles
N.W. Atlantic	*Nybelinia (Nybelinia) bisulcata*	viscera
N.W. Atlantic	*Nybelinia (Nybelinia) lamontae*	
Atlantic	*Nybelinia (Nybelinia) lingualis*	
Atlantic (Woods Hole)	*Otobothrium (Otobothrium) crenacolle*	flesh and viscera
Atlantic	*Otobothrium (Pseudotobothrium) dispacum*	
N.W. Atlantic	*Phyllobothrium loliginis*	stomachs
N.W. Atlantic	*Scolex pleuronectis*	intestine
N.W. Atlantic	*Tentacularia bicolor*	stomach
Atlantic	*Tentacularia coryphaena*	
Copepods (Silas and Ummerkutty 1967)		
N.W. Atlantic	*Brachiella ramosa*	gills
	Brachiella thynni	gills
N.W. Atlantic	*Caligus chelifer*	body surface
Atlantic, Mediterranean	*Caligus elongatus (C. rapax)*[1]	body surface
Atlantic ?	*Chondracanthus xiphiae*	gills
N.W. Atlantic	*Gloiopotes ornatus*	body surface

[1]Bassett-Smith, P. W. 1899. A systematic description of parasitic Copepoda found on fishes, with an enumeration of the known species. Proc. Zool. Soc. Lond. 1899:438-507. Incorrectly reported in Silas and Ummerkutty (1967) and reported as synonym of *rapax* in 1899 by Bassett-Smith.

bunova 1969; manuscript not seen by authors), the periods of intensive feeding coincide in time with the periods of greatest frequency of larvae caught in sampling nets. Juvenile swordfish 8 mm long will swallow fish that are as long as themselves (Tån-ning 1955).

Adult swordfish are opportunistic feeders, known to forage for their food from the bottom to the surface over great depths and distances. Their diet varies with location and species. According to Beardsley (see footnote 6), "Swordfish are diurnal feeders, rising to the surface and near surface waters at night. Over deep water they feed primarily on pelagic fishes and squids, while in shallower water large adults make feeding excursions to the bottom where the temperatures may be 5-10 °C and feed on demersal species."

In temperate waters of the Atlantic and Pacific, swordfish frequently bask on the surface. This behavior is rarely observed in tropical waters and is thought to facilitate digestion in temperate waters (surface waters being relatively warmer). Stomachs sampled from swordfish caught at the surface were either full or completely empty (Tibbo et al. 1961). They noted that swordfish would on occasion regurgitate everything from their stomachs before capture and sometimes would even evert the stomach.

There is some question as to the use of the sword in obtaining food. Goode (1883) reported that swordfish rise beneath a school of fish, striking to the right and left with their swords until they have killed a number of fish, which they then proceed to devour. Recent researchers (Tibbo et al. 1961) have found evidence from stomach contents that the swordfish uses its sword to kill some of its food. Scott and Tibbo (1968) stated the swordfish differs from the spearfishes (marlins and sailfish) in that the sword is long and dorsoventrally compressed. Thus, the swordfish appears to be more highly specialized for lateral slashing. They believed such a specialization would be pointless unless directed towards a vertically oriented prey or unless the swordfish slashes mainly while vertically oriented, as when ascending or descending.

3.42 Food

Arata (1954) examined stomach contents from larvae from the western Atlantic ranging in size from 7.8 to 192.1 mm. Only the two smallest specimens (7.8 and 9.0 mm) contained zooplankton, while for all of the other specimens, fish larvae were the main food source.

In the Pacific, swordfish at 9.0-14.0 mm feed on organisms such as Mysida, Phyllopoda, and Amphipoda and do not begin to feed on other fish until about 21 mm long (Yabe et al. 1959).

As juveniles, swordfish feed on squids, fishes, and some pelagic crustaceans. It is widely accepted that, in general, large predatory fishes eat whatever is available in the greatest abundance in their immediate environment (Scott and Tibbo 1968), and swordfish appear to be no exception. The major portion of their diet consists of squids, fishes, and occasional crustaceans and varies with location and species available (Table 2). There is no information available on either sexual differences in food habits or feeding habits in relation to size and sex.

3.43 Growth rate

Information on swordfish growth is limited and somewhat contradictory. Swordfish hatch at a length of 4.0-4.2 mm and larvae 5.5 mm are about 5 days old (Sanzo 1922). Arata (1954) stated that swordfish larvae have a high growth rate, about 0.6 mm/day, while Tibbo and Lauzier (1969) indicated that the growth rate was around 2 mm/day. In the Pacific, swordfish grow to 500-600 mm in the first year (Yabe et al. 1959).

In the northwest Atlantic, adults enter temperate waters in June, are usually thin, but will add 22-34 kg as the season advances (Tibbo et al. 1961). Beckett (1974) suggested that the growth of female swordfish may be rapid with a general age weight relationship of:

Age 1 -   4 kg
Age 2 -   15 kg
Age 3 -   40 kg
Age 4 -   70 kg
Age 5 - 110 kg

based on a rough analysis of size frequencies from commercial catches and the analysis of tagging results. However, Guitart-Manday (1964) believed that a swordfish of 160 cm and 59 kg in the southwest Atlantic (off Cuba) was 2 yr old.

In the Pacific, Yabe et al. (1959) indicated that swordfish 50-60 cm long were 1 yr old, that the fish caught commercially were predominately 4-5 yr old, and that swordfish spawned for the first time at 5-6 yr. Swordfish in the western Pacific grow about 25 cm/yr (Yabe et al. 1959), while eastern Pacific swordfish between 62 and 165 cm grow about 38 cm/yr (Kume and Joseph 1969).

There is good evidence for differential growth between males and females with females attaining the larger size (Cavaliere 1963; Guitart-Manday 1964; Kume and Joseph 1969). Females grow more rapidly than males and not only grow to a greater length than males, but are proportionally heavier at the same length (Skillman and Yong 1974). Beckett (1974) suggested that few males exceed 200 cm FL (fork length) (approximately 120 kg).

3.5 Behavior

3.51 Migrations and local movements

There are few data on migrations of swordfish. Tag return data from the northwest Atlantic indicate that swordfish either make very limited local movements during the year or return each year to the same feeding grounds (Beckett 1971[1]).

[1]Beckett, J. S. 1971. Canadian swordfish longline fishery. Working paper (SCRS/71/36) submitted to the Standing Committee on Research and Statistics, November 1971. Int. Comm. Conserv. Atl. Tunas, General Mola 17, Madrid 1, Spain.

Table 2.—Food organisms found in stomachs of adult swordfish by location (and reference).

N.W. Atlantic	Tetraodon
(Scott and Tibbo 1968)	Gempylus
Mackerel	Trichiurus
Barracudinas	Carcharhinus
Silver hake	Mediterranean
Redfish	(Cavaliere 1963, Part II)
Herring	Illex coindetii
Saury	Loligo todarus
Hake	Todarodes sagittatus
Snake mackerels	Engraulis encrasicholus
Filefish - triggerfish	Sardinella aurita
Lanternfish	Sardina pilchardus
Lancetfish	Scomberesox saurus
Bigeye scad	Anguilla anguilla
Sand lances	Boops boops
Viperfish	Lepidopus caudatus
Marlin-spike	Sea of Marmara (Artüz 1963)
Eels	Sinaris alcedo
Cuba (Guitart-Manday 1964)	Scomber scombrus
Brama brama	Scomber collas
Squids	Engraulis
Unidentified fishes	Merluccius sp.
Shrimp	Belone belone
Epinephelus guttatus	Unidentified fish
Unidentified crustacea	Ommastrephes (squids most
Octopus sp.	important)
Thysanoteuthis rhombus	Shrimps
Brazil (Ovchinnikov 1970)	N.W. Pacific (Yabe et al. 1959)
Teuthoidae	Lepidotidae
Vomer sp.	Cephalopoda
Octopoda (Cephalopoda)	Squid
Exocoetidae	Unidentified fish
E. Atlantic (Ovchinnikov 1970)	Acinaceidae
Cephalopods	Octopus
Peristedion	Cololabis saira
Brama	Amphipoda
Lepidopus	Shellfish
Thunnus	Chiasmodontidae
Trachipterus	Pacific (Nakamura 1949)
Palinurichthys	Cod
Coryphaena	Sebastodes
Alepisaurus	Myctophids

Although only 20 recoveries were recorded from 231 fish released between 1961 and 1976, all were recaptured within a few hundred kilometers of the release point. Only one long distance recovery has been recorded. Casey (see footnote 3) reported the recapture of a tagged swordfish on Georges Bank in August 1977 that had been released in the northern Gulf of Mexico in March 1974.

Tibbo et al. (1961) proposed two hypotheses on the migration of swordfish in the northwest Atlantic: 1) swordfish migrate to the north and east along the edge of the continental shelf during summer and return to the south and west in autumn, or 2) there are different groups of swordfish migrating from deep waters towards the continental shelf in summer and then move off to deep water again in autumn.

Carey and Roberson (1977[4]) conducted sonic tagging experiments in the eastern Pacific and western Atlantic which showed that most tagged swordfish generally stay inshore near the bottom during the day. At dusk they head seaward, swimming up and down through a considerable range of depths. After sunset they feed near the surface (2-13 m), and at sunrise they return to the 90-125 m inshore depth. Large individuals did

[4]Carey, F. G., and B. Roberson. 1977. Tracking swordfish in the Sea of Cortez. Hubbs-Sea World Res. Inst., Currents No. 1, p. 1-7. Hubbs-Sea World Research Institute, 1700 S. Shores Road, San Diego, CA 92109.

not come inshore but set up a meandering course seaward. These experiments involved two tagged fish in the Atlantic and five tagged fish in the Pacific.

In the eastern Pacific, catch records indicate a movement of fish from off the tip of Baja California during the spring and towards the north during the summer and fall (Kume and Joseph 1969). Kume and Joseph also suggested that swordfish along the coastal regions of South America move northward from Chile to Peru from June to September, and they further postulated that swordfish move seaward to spawn from November through February.

3.52 Schooling

Nothing found in the literature.

3.53 Pugnacity

Swordfish have a reputation for being a pugnacious fish. There are records of attacks on boats (Gray 1871; Smith 1956), whales (Brown 1960; Machida 1970; Peers and Karlsson 1976), and even submersibles (Zarudski 1967).

4 POPULATION

4.1 Structure

4.11 Sex ratio

In the northwestern Atlantic harpoon fishery, only large (120 kg) females are caught; however, in the longline fishery both sexes appear in the catch (Lee 1942; Tibbo et al. 1961). Guitart-Manday (1964) found both sexes in the catch in the southwestern Atlantic with males in greater numbers than females (72-28%); however, most of the large (75-137 kg) fish were females. Beckett (1974) found that sex ratios differ with temperature, as few males are found in the colder (under 18°C) water. In the Caribbean and adjacent regions, for example, Beckett found that males comprise from 67 to 100% of the catch.

In the eastern Pacific, the proportion of females to males in a sample of 1,449 swordfish was roughly equal over the size range 130-170 cm, but above this range the proportion of females became progressively higher (Kume and Joseph 1969).

4.12 Age composition

Using Beckett's (1974) age estimates and length-frequency data from Beardsley et al. (1979[9]), the age composition of swordfish in the western North Atlantic sport fishery is primarily ages three to five with some 1 and 2 yr olds and some 6 yr and older also being caught. In the Sea of Marmara (using Beckett's age estimates), commercially caught swordfish are primarily 3 and 4 yr olds with very few 2 yr olds and some 5 and 6 yr olds in the catch (Artüz 1963).

Yabe et al. (1959) concluded from an 8 yr study of the Pacific swordfish fishery that swordfish taken commercially in the North Pacific fishing grounds are approximately 2 yr old

[9]Beardsley, G. L., R. J. Conser, A. M. Lopez, M. Brassfield, and D. McClellan. 1979. Length and weight data for western Atlantic swordfish, *Xiphias gladius*. Coll. Vol. Sci. Pap., VIII(SCRS-1978):490-495. · Int. Comm. Conserv. Atl. Tunas, General Mola 17, Madrid 1, Spain.

and older, and the predominant age group consists of 4-5 yr olds, although there are some 1 yr olds caught.

4.13 Size composition

Swordfish grow to a very large size, occasionally attaining weights of over 500 kg. The world record swordfish taken by sportfishing gear was captured off Chile in 1953 and weighed 536 kg (International Game Fish Association 1979). Beckett (1974) reported a swordfish landed at Cape Breton, Nova Scotia, that weighed approximately 550 kg.

Size data from the Japanese longline fishery in the Atlantic show a broad range from 80 to 300 cm in length (rear of the orbit to the caudal fork) with the majority between 130 and 230 cm (Fig. 6). Beckett (see footnote 7) showed a rapid decline in the average size of swordfish caught in the Canadian longline fishery, from 120 kg round weight in 1963 to 60 kg in 1969. Part of this decline Beckett attributed to a gradual expansion of the fishery into warmer waters where smaller males are more common in the catches.

In the eastern Pacific, Kume and Joseph (1969) presented size data for swordfish taken by longline vessels (Fig. 7). The range was from 50 to 275 cm eye-fork length with the mode located at about 170 cm. In the western pacific, Yabe et al. (1959) presented length data for swordfish captured by the longline fishery from 1948 to 1956. Average size decreased steadily from about 170 cm to about 130 cm body length during that period (Fig. 8).

Length-weight relationships of swordfish are summarized in Table 3.

4.2 Abundance and density (of population)

4.21 Average abundance

No estimates of population size are available.

4.22 Changes in abundance

See section 4.24.

4.23 Average density

Nothing found in the literature.

4.24 Changes in density

In the western North Atlantic, catch per unit effort (CPUE) in the Canadian longline fishery declined from 2.88 fish/100 hooks in 1963 to 0.92 fish/100 hooks in 1965. This decline in CPUE was accompanied by a decrease in average size from 120 kg round weight to <60 kg (Beardsley see footnote 6). Part of this decline in average size may have been due to an expansion of the fishery to more southern grounds. In 1975, however, following a 4 yr period when there was no fishing due to restrictions on the sale of swordfish due to mercury contamination, CPUE had risen to 2.31 fish/100 hooks (Caddy 1976). Guitart-Manday (1975) showed a general increasing trend in the catch per days fishing for swordfish off Cuba by the Cuban longline fleet. CPUE in the Japanese longline fishery for the total Atlantic increased steadily from 1956 to 1968 then stabilized through 1975. (Fig. 9).

In the eastern Pacific, catch rates for swordfish by the

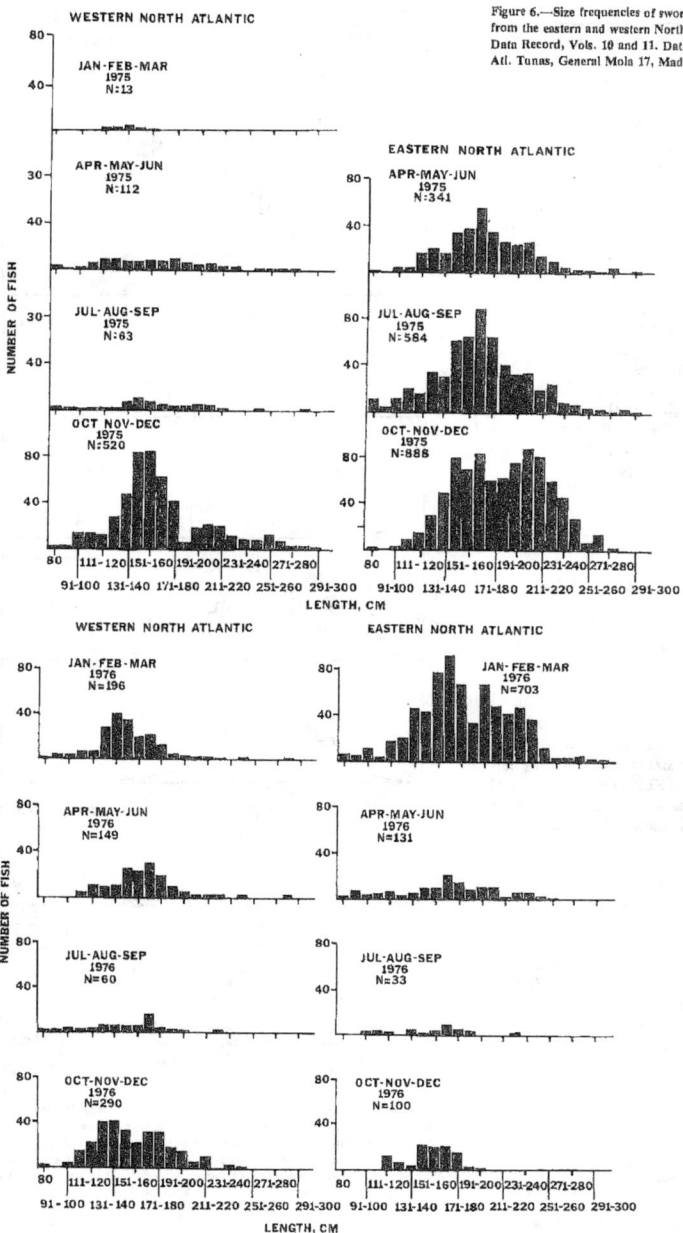

Figure 6.—Size frequencies of swordfish captured in the Japanese longline fishery from the eastern and western North Atlantic in 1975 and 1976 (data from ICCAT Data Record, Vols. 10 and 11. Data records available from Int. Comm. Conserv. Atl. Tunas, General Mola 17, Madrid 1, Spain).

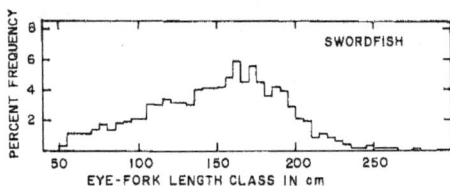

Figure 7.—Size-frequency curve, in percentage, for 1,449 swordfish caught by longline vessels in the eastern Pacific (from Kume and Joseph 1969, fig. 14 and caption).

Japanese longline fishery show generally an increasing trend in the two main fishing areas through 1969 (Fig. 10) (Joseph et al. 1974). The authors pointed out, however, that the increase may be due to an increase in the number of night sets and concentration on the more productive swordfish areas. Sakagawa and Bell (1978[10]) showed the same general increase in CPUE through 1969 but then a sharp decline through 1975 (Fig. 11, Area 3). CPUE over the entire Pacific, however, shows a gradual decline from 1958 through 1967 then stabilizing through 1975 (Fig. 12).

4.3 Natality and recruitment

4.31 Reproduction rates

See section 3.1.

4.32 Factors affecting reproduction

Nothing found in the literature.

4.33 Recruitment

A rough estimate of size at first capture using a comparison of commercial size frequencies by longline and harpoon in the mid-1960's indicates that size at first capture by harpoon is in the vicinity of 36 kg (155 cm FL) and probably slightly lower than this in the Pacific (Beardsley see footnote 6). The size at first capture by longline in the Atlantic is in the vicinity of 4.5 kg (80 cm FL)

4.4 Mortality and morbidity

4.41 Mortality rates

For western North Atlantic stocks, estimates of total mortality (Z) have been made for the harpoon fishery ($Z = 0.12 - 0.65$) and for the longline fishery ($Z = 0.16 - 0.59$). The natural mortality of swordfish may be relatively low when compared to other billfishes because of their longevity, and preliminary estimates derived from the relationship between M and K,

$$M = -0.0195 + 1.9388 \ (K),$$

[10]Sakagawa, G. T., and R. R. Bell. 1978. Swordfish, *Xiphias gladius*. *In* R. S. Shomura (editor), Summary report of the Billfish Stock Assessment Workshop, Pacific Resources, Honolulu, Hawaii, 5-14 December 1977. Mimeo. Rep., p. 43-55. Honolulu Laboratory, Southwest Fisheries Center, NMFS, NOAA, P.O. Box 3830, Honolulu, HI 96812.

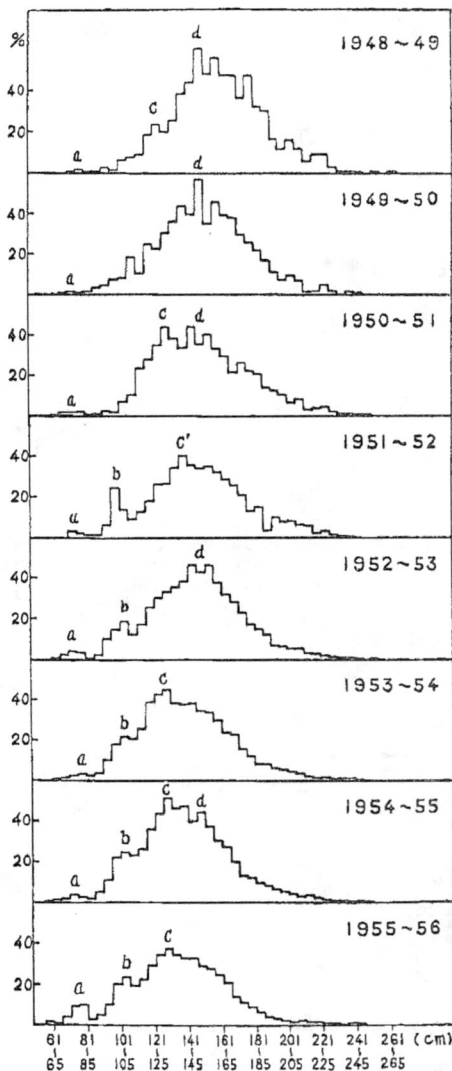

Figure 8.—Body length (distance from the posterior end of the upper jaw to the terminal of the hypural bone) compositions of swordfish from the North Pacific fishing grounds (140°-160°E) for an eight-year period from 1948-56 (from Yabe et al. 1959, fig. 25 and caption; authors note: the abscissa is body length in cm; although the ordinate is labeled percent, we do not believe the numbers as shown on the figure represent percent. The letters a, b, c, and d represent modes).

266

Table 3.—Length-weight relationships of swordfish, *Xiphias gladius*, sexes combined. The form of the equation is $y = ax^b$. Beardsley et al. (1979) used the geometric mean form; all others are predictive.

Author (area)	Measurement	Length range	N	a	b
Guitart-Manday (1964) (western Atlantic)	lower jaw fork length (cm) round weight (kg)	84.5-254.5	242	4.8643×10^{-7}	3.64237
Skillman and Yong (1974) (Pacific)	tip of upper bill to fork (cm) round weight (kg)	ca. 150-325	7	2.3296×10^{-7}	3.5305
Caddy (1976) (western Atlantic)	lower jaw fork length (cm) dressed weight (kg) ca 77% of round weight	ca. 50-260	---	1.30978×10^{-5}	3.0992
Beardsley et al. (1979)[1] (western Atlantic)	lower jaw fork length (cm) round weight (kg)	81-281	166	3.689×10^{-5}	3.2994
Rey and Gonzales-Garcés (1979)[2] (eastern Atlantic)	lower jaw fork length (cm) eviscerated weight (kg) (eviscerated weight = 0.75 × round weight[1.04])	90-234	486	5.17×10^{-6}	3.16
(Mediterranean)	lower jaw fork length (cm) eviscerated weight (kg)	94-180	105	9.7×10^{-7}	3.49
Amorim and Arfelli (1977) (western Atlantic)	eye-fork length (cm) gilled and gutted weight (kg) (gilled and gutted weight = 0.8009 × live weight[1.015])	---	1173	1.24×10^{-5}	3.04
Amorim (1977)[3] (western Atlantic)	eye-fork length (cm) gilled and gutted weight (kg)	73-240	865	5.36×10^{-6}	3.2

[1]See text footnote 9.
[2]See text footnote 11.
[3]Amorim, A. F. de. 1977. Informe preliminar sobre las investigaciones del pez espada *Xiphias gladius* en el sudeste sur del Brazil, en el período de 1971-76. Coll. Vol. Sci. Pap., VI(SCRS-1976):402-407. Int. Comm. Conserv. Atl. Tunas, General Mola 17, Madrid 1, Spain.

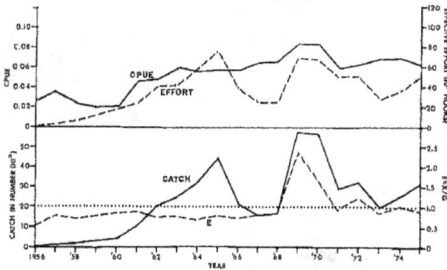

Figure 9.—Catch-per-unit-of-effort, effective number of hooks, catch in number, and index of effectiveness for Atlantic swordfish by the Japanese longline fishery, 1956-1975. X = effective number of hooks, G = nominal hooks (from Beardsley see text footnote 6, fig. 5 and caption).

indicate that *M* ranges from 0.21 to 0.43 (Beardsley see footnote 6).

4.42 Factors causing or affecting mortality

See section 3.34.

4.43 Factors affecting morbidity

See section 3.35.

4.44 Relations of morbidity to mortality rates

Nothing found in the literature.

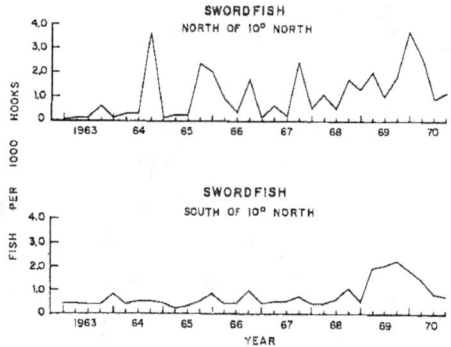

Figure 10.—Quarterly hook rate of swordfish expressed as catch in numbers per 1,000 hooks, for areas north and south of lat. 10°N in the eastern Pacific (from Joseph et al. 1974, fig. 19 and caption).

4.5 Dynamics of population (as a whole)

Nothing found in the literature.

4.6 The population in the community and the ecosystem

Studies on the distribution of swordfish worldwide and its relationship to the total ecosystem have not been done. However, the geographical distribution of swordfish in the northwestern Atlantic varies considerably due to marked

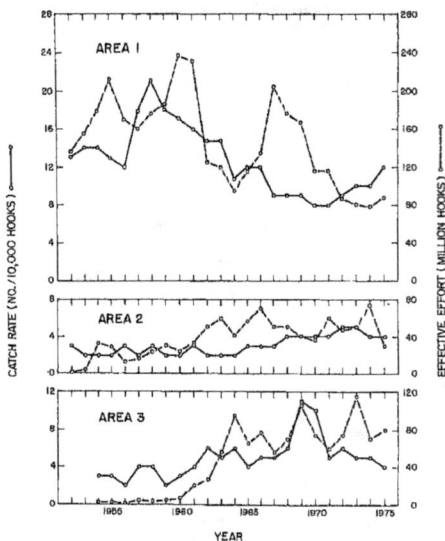

Figure 11.—Catch rates and effective fishing effort for swordfish in the Pacific (from Sakagawa and Bell see text footnote 10, fig. 6 and caption. Note: areas pertain to same areas portrayed in Fig. 2).

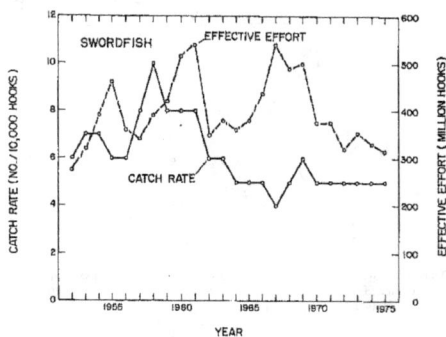

Figure 12.—Pacific-wide catch rate and effective fishing effort for swordfish (from Sakagawa and Bell, see text footnote 10, fig. 3 and caption).

seasonal variations in environmental conditions (Beckett 1974). In the eastern Pacific, swordfish populations are most abundant throughout the year in the inshore areas, probably in association with cool upwelled waters in that region (Kume and Joseph 1969).

Temperature apparently is important to the distribution of swordfish in all oceans (Yabe et al. 1959; Kume and Joseph 1969; Ovchinnikov 1970; Beckett 1974), and optimal surface temperatures appear to be 25°-29°C (Tåning 1955). Current systems such as the Gulf Stream in the Atlantic, the Kuroshio and Humboldt in the Pacific, as well as the Equatorials and Equatorial Countercurrents, play a major role in the distribution of swordfish because frontal zones with sharp gradients of temperature, salinity, and amounts of biogenous matter are created, and these are surrounded by areas of high productivity where swordfish and other predators concentrate (Ovchinnikov 1970).

5 EXPLOITATION

5.1 Fishing equipment

5.11 Gear

Swordfish are taken by longlines and harpoons in the commercial fishery. The harpoon for many years was the primary commercial gear in most areas of the Atlantic and Pacific. In the early 1960's, however, following the disclosure of substantial catches of swordfish by the Japanese tuna longline fishery and the Norwegian shark fishery in the Atlantic and high catch rates by exploratory fishing vessels in the northwest Atlantic using longline gear at night, most of the Canadian and United States vessels converted to longlines. In the eastern Atlantic and Mediterranean, longlines are used by the French and Spanish fleets, while harpoons are still the traditional gear of the Sicilian swordfish fleet.

In the Pacific, longlines and harpoons are used. In the eastern Pacific, however, the U.S. commercial fishery is restricted to the use of harpoons only.

Trolling and drift fishing using rod and reel gear are the primary fishing methods in the sport fishery. Until 1976, only trolling was used and involved initial visual observation of a swordfish basking at the surface before the lines were placed into the water. In 1976, however, sport fishermen off the east coast of Florida discovered that swordfish could be caught by drifting baited lines at night. Fishing success is substantially higher using this method than by the trolling method, and night sport fishing for swordfish now takes place all along the Atlantic and Gulf of Mexico coasts of the United States.

5.12 Boats

Commercial fishing vessels for swordfish range in size from the large, high-seas tuna longliners to small harpoon and longline boats < 10 m in length. Good descriptions of various types of fishing vessels used for longlining and harpooning for swordfish are described in Tibbo et al. (1961) and Guitart-Manday (1964). Vessels used in the sport fishery vary considerably in size and style. Rybovich (1965) presented a good description of a typical sport fishing vessel.

5.2 Fishing areas

5.21 General geographic distribution

Catch records from the high-seas tuna longline fishery indicate that swordfish are taken almost throughout the range of the fishery. For the most part, however, swordfish are incidental catches in the tuna longline fishery. Important directed commercial fisheries for swordfish are located in the western North Atlantic from the Grand and Georges Banks to

268

the Gulf of Mexico, in the eastern Atlantic and Mediterranean Sea, and in the South Atlantic off the coast of Brazil and Uruguay. In the western Pacific, the major fishing grounds extend from Japan eastward to long. 165°W.

Major sport fishing areas are located off the east coast of the United States from New York to Texas and off the coast of southern California. Some sport fishing for swordfish also takes place in the South Pacific, notably off Chile and Ecuador.

5.22 Depth ranges

Swordfish are caught by both commercial and sport fisheries over a wide range of depths. Generally, major fishing areas are located near continents in the Atlantic in depths of < 1,000 m. In the Pacific, major fisheries take place at considerably greater depths.

5.3 Fishing seasons

5.31 General patterns of seasons

Swordfish are caught throughout the year.

5.32 Dates of beginning, peak, and end of seasons

In the western Atlantic, the longline fishery for swordfish operates year round; however, there is a substantial geographic shift in the fishery with the seasons. In the summer and early fall, most fishing takes place off Grand and Georges Banks. In winter, many of the larger boats move into more southern waters, operating from Cape Hatteras southward around Florida and throughout the Gulf of Mexico. In the eastern Atlantic and Mediterranean, Artüz (1963) stated that landings from the Sea of Marmara peaked in May, while Rey and Gonzales-Garcés (1979[1]) presented data that showed the peak of the Spanish fishery during July to December.

In the eastern Pacific, longline catches are made all year long, but the peak season is from January through March (Kume and Joseph 1969). The California harpoon fishery operates mainly in summer and fall.

5.4 Fishing operations and results

5.41 Effort and intensity

In the Atlantic, effective fishing effort for swordfish by the Japanese longline fleet increased rapidly from 1956 to a peak in 1965 then fluctuated widely through 1975 (Fig. 9). The United States and Canadian fisheries expanded rapidly following the conversion to longline gear in 1962. In 1971, however, both fisheries were essentially terminated following restrictions on the sale of swordfish with high levels of mercury contamination in the flesh. In recent years, this fishery has undergone a resurgence and although no hard data are available, effort is probably as great if not greater than during the pre-1971 years.

The recreational fishery has historically been rather insignificant, both in terms of effort expended and in catch. In recent years, however, the development of a nighttime fishery

for swordfish has dramatically increased catches and this type of fishing has expanded all along the Gulf and Atlantic coasts of the United States.

In the Pacific, fishing effort by the Japanese longline fleet attained two major peaks, one in 1961 and the other in 1967, but since 1967 it has declined steadily through 1975 (Fig. 12). In the eastern Pacific, effort increased sharply from 1954 to 1964 then leveled off through 1975.

5.42 Selectivity

Nothing found in the literature.

5.43 Catches

Landings of swordfish from the Atlantic are shown in Table 4 and from the Pacific, Table 5.

6 PROTECTION AND MANAGEMENT

6.1 Regulatory (legislative measures)

6.11 Limitation or reduction of total catch

Restrictions on the sale of swordfish containing levels of mercury in the flesh > 0.5 ppm were imposed in Canada and the United States in the early 1970's. These restrictions caused the complete collapse of the Canadian fishery and severely reduced landings in the United States. Gradually, however, the U.S. fishery began to resume normal operations, but essentially in a clandestine fashion. In 1979, the mercury guidelines were raised to 1.0 ppm as a result of legal action initiated by the American Swordfish Association. By 1980, catch and fishing effort were probably at an all time high in the Atlantic and the fishery extended from Canada to Mexico.

6.12 Protection of portions of population

In Turkey, no swordfish < 10 kg can be caught, offered, stored, or exposed for sale (Artüz 1963).

6.2 Control or alteration of physical features of the environment

Nothing found in the literature.

6.3 Control or alteration of chemical features of the environment

Nothing found in the literature.

6.4 Control or alteration of the biological features of the environment.

Nothing found in the literature.

6.5 Artificial stocking

Nothing found in the literature.

7 POND FISH CULTURE

Not applicable.

[1]Rey, J. C., and A. Gonzales-Garcés. 1979. Nuevos datos sobre la pesquería Española de pez espada, *Xiphias gladius*, biología y morphometría. Coll. Vol. Sci. Pap.,VIII(SCRS-1978): 504-509. Int. Comm. Conserv. Atl. Tunas, General Mola 17, Madrid 1, Spain.

Table 4.—Landings in metric tons of swordfish, *Xiphias gladius*, from the Atlantic Ocean, 1967-1977. (ICCAT Statistical Bulletin, Vol. 8-1977, p. 25.)

Country	1967	1968	1969	1970	1971	1972	1973	1974	1975	1976	1977
Algeria	0	0	0	xx	xx	2	100	196	500	368	370
Argentina	100	300	500	400	100	100	48	0	10	111	132
Brazil	120	120	240	120	0	120	137	348	318	330	275
Bulgaria	0	0	0	0	0	0	0	0	0	0	3
Canada	4,800	4,400	4,300	4,800	0	0	0	0	21	15	113
Taiwan	0	0	0	0	0	750	1,092	821	928	935	922
Cuba	200	0	0	0	0	0	0	0	0	600	700
Cyprus	0	0	0	0	0	0	0	0	5	72	118
Ghana	0	0	0	0	0	0	0	0	0	0	642
Italy	1,900	1,400	2,000	1,800	2,900	3,700	2,700	1,500	1,500	2,140	1,935
Japan	754	1,121	2,273	3,175	1,576	1,805	998	1,369	1,500	809	792
Korea	0	0	0	0	0	0	0	0	451	1,147	1,240
Libya	300	500	xx	xx	100	xx	xx	xx	0	0	0
Malta	xx	xx	xx	100	200	200	200	171	191	156	199
Mexico	0	0	0	0	0	2	4	3	0	0	0
Morocco	204	240	270	231	360	273	201	211	133	198	151
Norway	300	200	600	400	200	xx	xx	xx	0	0	0
Panama	0	0	0	0	0	167	445	0	0	0	0
Poland	0	0	0	0	xx	0	100	0	0	0	0
Spain	3,390	4,550	4,600	4,060	4,484	4,510	4,938	3,593	3,836	2,905	3,976
Tunisia	0	0	0	xx	xx	xx	xx	5	0	0	0
Turkey	98	0	119	88	76	76	xx	6	0	0	0
USA	474	274	171	287	35	246	406	1,125	1,700	1,429	0
USSR	xx	xx	100	200	200	200	200	1,400	263	562	121
Venezuela	360	0	120	0	0	0	0	0	0	0	0
TOTAL:	13,000	13,105	15,293	15,661	10,231	12,151	11,569	10,748	11,356	11,777	11,689

Table 5.—Swordfish catches (metric tons) by countries for the Pacific Ocean (Sakagawa and Bell, see text footnote 10).

Year	Japan	Taiwan	Korea	United States	Chile	Peru	Others	Total
1952	11,182			157				11,339
1953	11,604			85				11,689
1954	13,301	77		14				13,392
1955	16,220	185		80				16,485
1956	12,167	254		163				12,584
1957	15,771	250		222				16,243
1958	20,815	247		279				21,341
1959	19,136	262		265				19,663
1960	22,944	273		192				23,409
1961	23,636	432		218				24,286
1962	14,037	544		23				14,604
1963	13,775	300		58				14,133
1964	9,703	300		109				10,112
1965	11,955	300		194	200	300		12,949
1966	13,283	600	41	277	200	200		14,601
1967	13,083	838	47	181	200	1,300		15,649
1968	12,983	974	55	118	200	800	100	15,230
1969	15,612	1,023	89	610	300	1,200	100	18,934
1970	11,301	1,053	115	558	200	2,400	100	15,727
1971	9,182	1,149	115	91	200	200	100	11,037
1972	8,846	1,111	115	157	100	600	100	11,029
1973	9,644	1,269	115	363	400	1,900	100	13,791
1974	9,517	1,157	115	384	218	270	3	11,664
1975	11,274	1,099	115	512	218	158		13,376

ACKNOWLEDGMENTS

We thank Donald de Sylva, Edward Houde, and Steven Berkeley of the University of Miami Rosenstiel School of Marine and Atmospheric Science, Shoji Ueyanagi of the Far Seas Fisheries Research Laboratory, Bruce B. Collette of the National Marine Fisheries Service Systematics Laboratory, and an anonymous reviewer for their constructive and thorough reviews of the manuscript.

LITERATURE CITED

AMORIM, A. F. de, and C. A. ARFELLI.
  1977. Contribução al conhecimento da biologia e pesca do espadarte e alguhões no litoral sul-sudeste do Brasil. 1 Congresso Paulista de Agronomia, 5-9 Setembro, Sao Paulo, Brasil.
ARATA, G. F., Jr.
  1954. A contribution to the life history of the swordfish, *Xiphias gladius* Linnaeus, from the South Atlantic coast of the United States and the Gulf of Mexico. Bull. Mar. Sci. Gulf Caribb. 4:183-243.
ARNOLD, E. L., Jr.
  1955. Notes on the capture of young sailfish and swordfish in the Gulf of Mexico. Copeia 1955:150-151.
ARTÜZ, M. I.
  1963. Contribution to the knowledge of the biology of the swordfish (*Xiphias gladius* L.) in the Sea of Marmara. Proc. Gen. Fish. Coun. Medit. 7:459-471. (Tech. Pap. 47.)
BECKETT, J. S.
  1974. Biology of swordfish, *Xiphias gladius* L., in the Northwest Atlantic Ocean. *In* R. S. Shomura and F. Williams (editors), Proceedings of the International Billfish Symposium, Kailua-Kona, Hawaii, 9-12 Aug. 1972. Part 2. Review and contributed papers, p. 105-106, U.S. Dep. Comm., NOAA Tech. Rep. NMFS SSRF-675.
BIGELOW, H. B., and W. C. SCHROEDER.
  1953. Fishes of the Gulf of Maine. U.S. Fish Wildl. Serv. Fish. Bull. 53, 577 p.
BLOCH, M. E., and J. G. SCHNEIDER.
  1801. Systema Ichthyologiae. Photographic facsimile, Wheldon and Wesley, Ltd., 1967, 584 p.
BOGOROV, V. G., and T. S. RASS.
  1961. On the productivity and prospects of fishing in waters of the Indian Ocean. [In Russ.] Okeanologiya 1:107-109.
BOZHKOV, A. T.
  1975. An instance of an attack by a swordfish (*Xiphias gladius*) on a mako shark (*Isurus glaucus*). [In Russ.] Voprosy Ikhtiol. 15(5):934-935. (Transl. J. Ichthyol. 15:842-843.)
BROWN, S. G.
  1960. Swordfish and whales. Nor. Hvalfangst-Tid. (The Norwegian Whaling Gazette) 49:345-351.
CADDY, J. F.
  1976. A review of some factors relevant to management of swordfish

fisheries in the northwest Atlantic. Can. Fish. Mar. Ser. Tech. Rep. 633:1-36.

AVALIERE, A.
1963. Studi sulla biologica e pesca di *Xiphias gladius* L. Nota II. Boll. Pesca Piscic. Idrobiol. 18:143-170. [Transl. Fish. Res. Board Can. Trans. Ser. 2298.]

HACKO, P. I., S. D. THOMAS, and C. M. PILLAY.
1967. Scombroid fisheries of Madras State, India. Proc. Symp. Scombroid Fishes, Part III. Mar. Biol. Assoc. India, Symp. Ser. 1:1006-1008.

HURCH, R. E.
1968. Broadbill swordfish in deep water. Sea Front. 14:246-249.

UVIER, G., and A. VALENCIENNES.
1831. Histoire naturelle des poissons. Paris Vol. 8, 509 p.

ERANIYAGALA, P. E. P. (editor)
1951. The Istiophoridae and Xiphiidae of Ceylon. Spolia Zeylanica, Bull. Nat. Mus. Ceylon 26:137-142.

e SYLVA, D. P.
1962. Red-water blooms off northern Chile, April-May 1956, with reference to the ecology of the swordfish and the striped marlin. Pac. Sci. 16:271-279.

UNCKER, G.
1936. Svaerdfisch (*Xiphias gladius* L.); i danske Farvande. Flora Fauna Kjobenh. 42:92-94.

ISH, M. P.
1926. Swordfish eggs. Bull. N.Y. Zool. Soc. 29:206-207.

ITCH, J. E.
1960. Swordfish. *In* (State Calif., Dep. Fish Game.) Calif. Fisheries Resources to the Year 1960, 79 p.

OWLER, H. W.
1928. The fishes of Oceania. Mem. Bernice P. Bishop Mus. 10, 540 p.

OODE, G. B.
1883. Materials for a history of the sword-fishes. Rep. U.S. Commer. Fish. (1880)8:287-394.

ORBUNOVA, N. N.
1969. Breeding grounds and food of the larvae of the swordfish [*Xiphias gladius* Linné (Pisces, Xiphiidae) sic]. Probl. Ichthyol. 9:375-387.

OSLINE, W. A.
1968. The suborders of perciform fishes. Proc. U.S. Natl. Mus. 124(3647):1-78.

RAY, J. E.
1871. On the injury inflicted on ships by the broad-finned swordfish of the Indian Ocean. Ann. Mag. Nat. Hist., Ser. 4, 8:338-339.

REENWOOD, P. H., D. E. ROSEN, S. H. WEITZMAN, and G. S. MYERS.
1966. Phyletic studies of teleostean fishes, with a provisional classification of living forms. Bull. Am. Mus. Nat. Hist. 131:339-455.

UITART-MANDAY, D.
1964. Biología pesquera del emperador o pez de espada, *Xiphias gladius* Linnaeus (Teleostomi: Xiphiidae) en las aguas de Cuba. [In Sp., Engl. synop.] Poeyana, Ser. B., No. 1, 37 p.
1975. Short-range marine pelagic fishing of northwest Cuba. Cuban Sci. Acad. Oceanogr., Inst. Oceanogr. Ser. 31. (Transl. avail. U.S. Dep. Commer., NOAA, NMFS TT-77-55012, 41 p.)

ÜNTHER, A. C. L. G.
1880. An introduction to the study of fishes. Adam and Charles Black, Edinburgh, 720 p.

ECTOR, J.
1875. Descriptions of five new species of fishes obtained in the New Zealand seas by H.M.S. "Challenger" Expedition, July 1874. Ann. Mag. Nat. Hist., Ser. 4, 15:78-82.

IO, J.-S.
1963. On five species of Formosan parasitic copepods belonging to the suborder Caligoida. Crustaceana 5:81-98.

LES, C.
1971. *Fistulicola plicatus* (Cestoda) and *Tristoma* spp. (Trematoda) on swordfish from the northwest Atlantic. J. Fish. Res. Board Can., 28:31-34.

INTERNATIONAL GAME FISH ASSOCIATION.
1979. World record game fishes. Int. Game Fish Assoc., Ft. Lauderdale, Fla., 272 p.

AKUCZUN, B.
1971. Swordfish (*Xiphias gladius* L.) *Xiphiidae* at the Wolin Island's Seashore (Western Poland). [In Polish, Engl. Summ.] Przegl. Zool. 15:297-298.

ONES, E. C.
1971. *Isistius brasiliensis*, a squaloid shark, the probable cause of crater wounds on fishes and cetaceans. Fish. Bull., U.S. 69:791-798.

JONES, S.
1958. Notes on eggs, larvae, and juveniles of fishes from Indian waters I. *Xiphias gladius* Linnaeus. Indian J. Fish. 5:357-361.

JOSEPH, J., W. L. KLAWE, and C. J. ORANGE.
1974. A review of the longline fishery for billfishes in the Eastern Pacific Ocean. *In* R. S. Shomura and F. Williams (editors), Proceedings of the International Billfish Symposium, Kailua-Kona, Hawaii, 9-12 Aug. 1972. Part 2. Review and contributed papers, p. 309-331. U.S. Dep. Comm., NOAA Tech. Rep. NMFS SSRF-675.

KONDRITSKAYA, S. I.
1970. The larvae of the swordfish [*Xiphias gladius* (L.)] from Mozambique Channel. J. Ichthyol. 10:853-854.

KUME, S., and J. JOSEPH.
1969. Size composition and sexual maturity of billfish caught by the Japanese longline fishery in the Pacific Ocean east of 130°W. [In Engl., Jpn. Summ.] Bull. Far. Seas. Fish. Res. Lab. (Shimizu) 2:115-162.

LaMONTE, F.
1944. Note on breeding grounds of blue marlin and swordfish off Cuba. Copeia 1944:258.

LaMONTE, F., and D. E. MARCY.
1941. Swordfish, sailfish, marlin, and spearfish. Ichthol. Contrib. Int. Game Fish Assoc., N.Y. 1(2):1-24.

LEACH, W. E.
1818. Description of a swordfish found in the Firth of Forth in June, 1811. Mem. Wernerian Nat. Hist. Soc. 2:58-60.

LEE, R. E.
1942. The occurrence of female sword-fish in southern New England waters, with a description of their reproductive condition. Copeia 1942:117-119.

LESHCHEVA.
1967. The food of the pelagic larvae of sea fishes under various conditions of existence. Author's abstract of thesis, Moscow. [In Russ., cited in Gorbunova 1969.]

LINNAEUS, C.
1758. Systema naturae. Regnum animale. 10th ed. Holmiae, 824 p. [Photographic facsimile printed by Br. Mus. Nat. Hist. 1956.]

LOBELL, M. J.
1947. The fisheries of Chile, present status and future possibilities (United States Fisheries Mission to Chile). U.S. Dep. Interior, Fish Wildl. Serv., p. 14-20.

LÜTKEN, C. F.
1880. Spolia Atlantica. Bidrad til Kundskab om Formforandringer hos Fiske under deres Vaext og Udvikling, saerligt hos nogle af Atlanterhavets Højsøfiske. [In Dan., Fr. summ.] Kgl. Danske Vidensk. Seksk. Skr. 5, Ser. 12:441-447, 592-593. [English translation of the French summary published as: Lütken, C. F. 1881. Spolia Atlantica: Contributions to the knowledge of the changes of form in fishes during their growth and development, especially in the pelagic fishes of the Atlantic. Ann. Mag. Nat. Hist. Ser. 5, 7:1-14, 107-123. English translation of Danish in Goode (1883).]

MACHIDA, S.
1970. A sword-fish sword found from a north Pacific Sei whale. Sci. Rep. Whale Res. Inst. 22:163-164.

MAKSIMOV, V. P.
1968. Brief communications. Swordfish attack on a shark. Prob. Ichthyol. 8:756.

MARKLE, G. E.
1974. Distribution of larval swordfish in the Northwest Atlantic Ocean. *In* R. S. Shomura and F. Williams (editors), Proceedings of the International Billfish Symposium, Kailua-Kona, Hawaii 9-12 Aug. 1972. Part 2. Review and contributed papers, p. 252-260. U.S. Dep. Comm., NOAA Tech. Rep. NMFS SSRF-675.

MATSUMOTO, W. M., and T. K. KAZAMA.
1974. Occurrence of young billfishes in the central Pacific Ocean. *In* R. S. Shomura and F. Williams (editors), Proceedings of the International Billfish Symposium, Kailua-Kona, Hawaii 9-12 Aug. 1972. Part 2. Review and contributed papers, p. 238-251. U.S. Dep. Comm., NOAA Tech. Rep. NMFS SSRF-675.

NAKAMURA, H.
1949. The tunas and their fisheries. Takeuchi Shōbo, Tokyo, 118 p. (Transl. by W. G. Van Campen, 1952, U. S. Fish Wildl. Serv., Spec. Sci. Rep. Fish. 82, 115 p.)

NAKAMURA, H., T. KAMIMURA, Y. YABUTA, A. SUDA, S. UEYANAGI, S. KIKAWA, M. HONMA, M. YUKINAWA, and S. MORIKAWA.
1951. Notes on the life-history of the sword-fish, *Xiphias gladius* Linnaeus. Jpn. J. Ichthyol. 1:264-271.

NAKAMURA, I.
    1974. Some aspects of the systematics and distribution of billfishes. In R. S. Shomura and F. Williams (editors), Proceedings of the International Billfish Symposium, Kailua-Kona, Hawaii, 9-12 Aug. 1972. Part 2. Review and contributed papers, p. 45-53. U. S. Dep. Comm., NOAA Tech. Rep. NMFS SSRF-675

NAKAMURA, I., T. IWAI, and K. MATSUBARA.
    1968. A review of the sailfish, spearfish, marlin, and swordfish of the world. [In Jpn.] Kyoto Univ., Misaki Mar. Biol. Ins. Spec. Rep. 4, 95 p.

NICHOLS, J. T.
    1923. Two new fishes from the Pacific Ocean. Am. Mus. Novit. 94, 3 p.

NISHIKAWA, Y., and S. UEYANAGI.
    1974. The distribution of the larvae of swordfish, Xiphias gladius, in the Indian and Pacific Oceans. In R. S. Shomura and F. Williams (editors), Proceedings of the International Billfish Symposium, Kailua-Kona, Hawaii, 9-12 Aug. 1972. Part 2. Review and contributed papers, p. 261-264. U.S. Dep. Comm., NOAA Tech. Rep. NMFS SSRF-675.

OSIPOV, V. G.
    1968. Some features of the distribution of tuna and other pelagic fishes in the northwestern Indian Ocean. Prob. Ichthyol. 8:22-28.

OVCHINNIKOV, V. V.
    1970. Mech-ryba i parusnikovye (Atlanticheskii okean. Ekologiya i funktsional 'naya morfologiya). (Swordfish and billfishes in the Atlantic Ocean. Ecology and functional morphology.) Nauch-issled. Inst. Ryb. Khoz. Okeanogr., Kaliningrad, 106 p. [Translated by Israel Prog. Sci. Transl., 77 p.; avail. U.S. Dep. Commer., Natl. Tech. Inf. Serv., Springfield, Va, as TT71-50011.]

PADOA, E.
    1956. Fauna e flora del Golfo di Napoli - 38. Monografia: Uova, larve e stadi giovanili di Teleostei. Divisione: Scombriformes (Famiglie Scombridae, Thunnidae, Gempylidae, Trichiuridae, Istiophoridae, Xiphiidae), p. 471-521. [Eggs, larvae and juveniles stages of the Scombriformes (in part) of the Gulf of Naples. (In Ital.) Translated by J. P. Wise and G. M. Ranallo, Transl. 12 Trop. Atl. Biol. Lab., Bur. Comm. Fish., Miami, Fla. 49 p.]

PEERS, B., and K. Ø. Karlsson.
    1976. Recovery of a swordfish (Xiphias gladius) sword from a fin whale (Balaenoptera physalus) killed off the west coast of Iceland. Can. Field-Nat. 90:492-493.

PENRITH, M. J., and D. L. CRAM.
    1974. The Cape of Good Hope: A hidden barrier to billfishes. In R. S. Shomura and F. Williams (editors), Proceedings of the International Billfish Symposium, Kailua-Kona, Hawaii, 9-12 Aug. 1972. Part 2. Review and contributed papers, p. 175-187. U.S. Dep. Comm., NOAA Tech. Rep. NMFS SSRF-675.

RASS, D. T. S.
    1965. The commercial fish fauna and fishery resources of the Indian Ocean. Tr. In-ta Okeanol. 80 AN SSSR (as cited in Osipov 1968).

REGAN, C. T.
    1909. On the anatomy and classification of the Scombroid fishes. Ann. Mag. Nat. Hist., Ser. 8, 3:66-75.
    1924. A young swordfish (Xiphias gladius), with a note on Chpeolabrus. Ann. Mag. Nat. Hist., Ser. 9, 13:224-225.

RICH, W. H.
    1947. The swordfish and the swordfishery of New England. Proc. Portland Soc. Nat. Hist. 4:1-102.

RICHARDS, W. J.
    1974. Evaluation of identification methods for young billfishes. In R. S. Shomura and F. Williams (editors), Proceedings of the International Billfish Symposium, Kailua-Kona, Hawaii, 9-12 Aug. 1972. Part 2. Review and contributed papers, p. 62-72. U.S. Dep. Comm., NOAA Tech Rep. NMFS SSRF-675.

ROSA, H.
    1950. Scientific and common names applied to tunas, mackerels and spearfishes of the world with notes on their geographic distribution. FAO, Wash., D.C., 235 p.

ROYCE, W. F.
    1957. Observations on the spearfishes of the central Pacific. U.S. Fish. Wildl. Serv., Fish. Bull. 57:497-554.

RYBOVICH, J.
    1965. Sportfishermen [boat]. In A. J. McClane (editor), McClane's Standard Fishing Encyclopedia, p. 851-862. Holt, Rinehart, and Winston, N.Y., 1058 p.

SANZO, L.
    1909. Uova e larve di Auxis bisus. Monit. Zool. Ital. 20:79-80. (Translated by W. G. Van Campen; avail. Natl. Mar. Fish. Serv., Honolulu, 3 p.)

1910. Uovo e larva di Pesce-spada (Xiphias gladius L.). Riv. mens. Pesca Idrobiol. 12:206-209.
    1922. Uova e larve di Xiphias gladius L. R. Com. Talassogr. Ital. Mem. 79,17 p.
    1930. Giovanissimo stadio larvale di Xiphias gladius L. di mm. 6.4. R. Com. Talassogr. Ital. Mem. 170, 8 p.

SCOTT, W. B., and S. N. TIBBO.
    1968. Food and feeding habits of swordfish, Xiphias gladius, in the western North Atlantic. J. Fish Res. Board Can. 25:903-919.

SELLA, M.
    1911. Contributo alla conoscenza della riproduzione e dello sviluppo del pesce spada (Xiphias gladius L.). R. Com. Talassogr. Ital. Mem. 2:1-16.

SERBETIS, K.
    1951. A new form of Xiphias thermaicus n. sp. [In Greek, Ital. summ.] Prakt. Akad. Athenon. 22:269-273.

SILAS, E. G.
    1967. Parasites of scombroid fishes. Part I. Monogenetic trematodes, digenetic trematodes, and cestodes. Proc. Symp. Scombroid Fishes, Part 3. Mar. Biol. Assoc. India, Symp. Ser. 1:799-875.

SILAS, E. G., and A. N. P. UMMERKUTTY.
    1967. Parasites of scombroid fishes. Part II. Parasitic Copepoda. Proc. Symp. Scombroid Fishes, Part 3. Mar. Biol. Assoc. India, Symp. Ser. 1:876-993.

SKILLMAN, R. A., and M. Y. Y. YONG.
    1974. Length-weight relationships for six species of billfishes in the central Pacific Ocean. In R. S. Shomura and F. Williams (editors), Proceedings of the International Billfish Symposium, Kailua-Kona, Hawaii, 9-12 Aug. 1972. Part 2. Review and contributed papers, p. 126-137. U.S. Dep. Comm., NOAA Tech. Rep. NMFS SSRF-675.

SMITH, J. L. B.
    1956. Pugnacity of marlins and swordfish. Nature (Lond.) 178:1065.

STARK, W.
    1960. Spear of swordfish, Xiphias gladius Linnaeus, imbedded in a silk shark, Eulamia floridana (Schroeder and Springer). Q. J. Fla. Acad. Sci. 23:165-166.

STEINDACHNER, F.
    1868. Übersicht der Meeresfische an den Küsten Spanien's und Portugal's. Fam. Xiphiidae. Sitzber. Akad. Wiss. Wien 57 (part 1):396-398.

STRASBURG, D. W.
    1970. A report on the billfishes of the central Pacific Ocean. Bull. Mar. Sci. 20:575-604.

TÅNING, Å. V.
    1955. On the breeding areas of the swordfish (Xiphias). Pap. Mar. Biol. Oceanogr., Deep Sea Res., suppl. to vol. 3:438-450.

TIBBO, S. N., L. R. DAY, and W. F. DOUCET.
    1961. The swordfish (Xiphias gladius L.), its life-history and economic importance in the northwest Atlantic. Fish. Res. Board Can., Bull. 130, 47 p.

TIBBO, S. N., and L. M. LAUZIER.
    1969. Larval swordfish (Xiphias gladius) from three localities in the western Atlantic. J. Fish. Res. Board Can. 26:3248-3251.

TSI-GEN, S.
    1960. Larvae and juveniles of tunas, sailfishes, and swordfish (Thunnidae, Istiophoridae, Xiphiidae) from the central and western part of the Pacific Ocean. Trudy Okean. Inst. Akad. Nauk SSSR 41:175-191. [In Russ., transl. avail. Inter-American Tropical Tuna Commission, Scripps Institution of Oceanography, La Jolla, CA 92037.]

UCHIYAMA, J. H., and R. S. SHOMURA.
    1974. Maturation and fecundity of swordfish, Xiphias gladius, from Hawaiian waters. In R. S. Shomura and F. Williams (editors), Proceedings of the International Billfish Symposium, Kailua-Kona, Hawaii, 9-12 Aug. 1972. Part 2. Review and contributed papers, p. 142-148. U.S. Dep. Comm., NOAA Tech. Rep. NMFS SSRF 675.

WEBB, B. F.
    1972. Broadbilled swordfish from Tasman Bay, New Zealand. N.Z. J. Mar. Freshwater Res. 6:206-207.

WISE, J. P., and C. W. DAVIS.
    1973. Seasonal distribution of tunas and billfishes in the Atlantic. U.S. Dep. Commer., NOAA NMFS Tech. Rep. SSRF-662, 24 p.

YABE, H.
    1951. Larva of the swordfish Xiphias gladius. [In Jpn., Engl. summ.]. Jpn. J. Ichthyol. 1:260-263.

YABE, H., S. UEYANAGI, S. KIKAWA, and H. WATANABE.
    1959. Study on the life history of the sword-fish, Xiphias gladius

Linnaeus. [In Jpn., Engl. summ.] Rep. Nankai Reg. Fish. Res. Lab.
10:107-150. [Translation by Masaru Fujiya available.]

YAMAGUTI, S.

1968a. Monogenetic trematodes of Hawaiian fishes. Univ. Hawaii
Press, Honolulu, 287 p.

1968b. Cestode parasites of Hawaiian fishes. Pac. Sci. 22:21-36.

1970. Digenetic trematodes of Hawaiian fishes. Keigaku Pub. Co.,
Tokyo, 436 p.

YASUDA, F., H. KOHNO, A. YATSU, H. IDA, P. ARENA, F. LI GRECI, and
Y. TAKI.

1978. Embryonic and early larval stages of the swordfish, *Xiphias
gladius*, from the Mediterranean. J. Tokyo Univ. Fish. 65:91-97.

ZARUDZKI, E. F. K.

1967. Swordfish rams the 'Alvin'. Oceanus 13(4):14-18.

21

☆ U. S. GOVERNMENT PRINTING OFFICE: 1981—593-351/15 REGION 10

Index

www.ingramcontent.com/pod-product-compliance
Lightning Source LLC
Chambersburg PA
CBHW061002280326
41935CB00009B/806